REMARKABLE CITIES AND THE FIGHT AGAINST CLIMATE CHANGE: 43 RECOMMENDATIONS TO REDUCE GREENHOUSE GASES AND THE COMMUNITIES THAT ADOPTED THEM

SUSTAINABLE DEVELOPMENT CODE: CLIMATE CHANGE CHAPTER

Jonathan Rosenbloom

ENVIRONMENTAL LAW INSTITUTE
Washington, D.C.

Contents

About the Author

Jonathan Rosenbloom is the Dwight D. Opperman Distinguished Professor of Law at Drake Law School. His scholarship explores issues relevant to local governments and sustainability, with a particular focus on land use. He is a former U.S. Circuit Court clerk, attorney for the federal government and a large law firm, and commissioner on the Des Moines Plan and Zoning Commission. He is also the founding director of the Sustainable Development Code, a model land use code providing local governments with the best sustainability practices in land use. Jonathan  has degrees from the Rhode Island School of Design, New York Law School, and Harvard Law School. He is happiest spending time with his wife and daughters and working with wood.

Acknowledgments

Getting the Sustainable Development Code (SDC) to this point has been an ongoing and gargantuan task for almost 10 years. The SDC is truly an interdisciplinary collaboration among academics and practicing attorneys, planners, architects, developers, local staff, and others devoted to improving the lives of all people. Drake University Law School has spearheaded this project in collaboration with almost a dozen law and planning schools and dozens of individuals. Without the drive and vision of many practitioners, academics, and students, this project would be impossible.

Each of the 43 recomendations recognizes the authors and editors who contributed to drafting that recommendation. Others who have been instrumental to the SDC and the Climate Change chapter include Drake University Law School Dean Jerry Anderson, Don Elliot, Jim Hansen, John Lorentzen, John Mitola, and Justin Platts. A special thanks to Chris Duerksen, who not only co-edited most of the briefs included in this book and chairs the SDC Advisory Committee, but also has been the moral compass of the SDC for over a decade. We also thank all the law and planning students that are named in this book, especially SDC 2018 Summer Fellows Tyler Adams, Brandon Hansom, Alec LeSher, and Kyler Massner. We thank the entire team at the Environmental Law Institute for making this possible.

Finally, without the financial support of Musco Lighting, Juhl Energy, and John Lorentzen, the SDC would simply not have been possible. In addition, RDG Planning and Design, P7 Design, and Clarion Associates LLP have financially supported the SDC to help make the Climate Change chapter a reality. Thank you.

—Jonathan Rosenbloom

Preface

In *Invisible Cities*, Italo Calvino tells the fictitious story of secret meetings between Marco Polo and Kublai Khan. Each night Polo describes the wonders found in cities in Khan's empire. Khan ultimately figures out that his empire is in far worse condition than Polo describes. Khan tells Polo that he knows his empire is crumbling and he and his people are in a downward spiral. Polo replies:

> Yes, the empire is sick, and what is worse, it is trying to become accustomed to its sores. This is the aim of my explorations: examining the traces of happiness still to be glimpsed, I gauge its short supply. If you want to know how much darkness there is around you, you must sharpen your eyes, peering at the faint lights in the distance.[1]

As the federal government pulls back from climate regulation and turns a blind eye to climate science and local plight, it is cities that provide a faint light and hope for the future. And yet, local governments face an uncertain and daunting future. Societal, environmental, political, and economic changes test the survival of many communities. These changes include higher obesity rates,[2] disparities in economic equality not seen since before the Great Depression,[3] state and federal hostility to local action,[4] increase in the sharing economy and autonomous vehicles,[5] and, potentially most distressing, a warmer climate[6] and irreplaceable loss of biodiversity.[7]

Paralleling these changes is an explosion of development that will rival post-World War II land use expansion. The U.S. population is projected to increase by almost 70 million people by 2040.[8] This increase and the phasing out of older buildings will require massive amounts of development, including approximately 90 billion additional square feet of commercial, retail, and industrial space and 80 million new residential units.[9]

If development patterns in the next 20-30 years replicate development patterns for the last 20-30 years, 40 million undeveloped acres will be destroyed (approximately the size of Oklahoma).[10] Communities across the country will lose critical ecosystems and habitats (forests, prairies, wetlands, etc.). Such a loss will add stress to already overstressed natural and man-made systems and will increase natural hazard risks to people, ecosystems, and communities.

Most communities' development codes were not designed for this on-coming wave of development and uncertainty. They were designed to address a static environment, society, and economy. In most part, they were not written to confront the kind of dynamic and unpredictable systems local communities face today.

Our current understanding of climatic and other changes suggests that continuing to allow development pursuant to existing codes may result in catastrophic losses across the social, economic and environmental spectrums. Many local governments, however, do not have the time or resources to research amendments to their development codes to address the myriad of changes and uncertainty they face.

The Sustainable Development Code was created to fill this gap and introduce additional community control.

The Sustainable Development Code (SDC)

What is and should be the role of development codes in planning for uncertainty? The SDC addresses this question by rethinking development so that "development" and "growth" help build sustainable communities and resilience. The SDC ensures that communities grow in a way that harmonizes development with ecosystems, nature, economic equality, and other core values that make communities . . . communities.

Officially launched on May 15, 2019, the SDC, www.sustainablecitycode. org, "aims to help all local governments, regardless of size and budget, build more resilient, environmentally conscious, economically secure and socially equitable communities."[11] The SDC provides local governments with best practices for particular issues with a focus on development codes and the development review process.

The SDC researches and identifies best-practices as adopted in local ordinances across the U.S. and across a broad spectrum of sustainability issues. Through a rigorous editorial and interdisciplinary process, the SDC summarizes this research and provides concrete ways for communities to amend development codes and adapt to changes as they occur.[12]

As of December 1, 2019, the SDC consisted of over 350 recommendations across 32 chapters. The 32 chapters, set forth in the textbox below, correspond to 32 areas where development implicates issues relevant to sustainability. The chapters serve as a menu for communities to identify the issues that they are confronting.

SDC Subchapters

Environmental Health & Natural Resources
1.1 Climate Change
1.2 Low-Impact Development & Stormwater Management
1.3 Sensitive Lands & Wildlife Habitat
1.4 Water Supply Quality & Quantity
1.5 Water Conservation
1.6 Solid Waste Management & Recycling
1.7 Urban Forestry & Vegetation

Natural Hazards
2.1 Floodplain & River Corridor Land Use
2.2 Wildfire Hazards and the Wildland-Urban Interface
2.3 Coastal Hazards
2.4 Steep Slope Hazards
2.5 Hazard Mitigation & Resiliency

Land Use & Community Character
3.1 Development Patterns & Infill
3.2 Development Densities
3.3 Mixed-Use
3.4 Transit-Oriented Development
3.5 Historic Preservation & Adaptive Reuse
3.6 Parking

Mobility & Transportation
4.1 Complete Streets/Safe Streets
4.2 Bicycle Mobility
4.3 Pedestrian Mobility
4.4 Public Transit
4.5 Autonomous Vehicles & New Technology

Community
5.1 Housing Affordability
5.2 Housing Diversity

Healthy Neighborhoods, Housing, & Food Security
6.1 Community Health & Safety
6.2 Food Security & Sovereignty

Energy
7.1 Wind Energy
7.2 Solar Energy
7.3 Other Energy Generation Systems
7.4 District Energy Systems
7.5 Energy Conservation & Efficiency

Each chapter consists of 30-40 recommendations. Each recommendation is sorted into one of three categories: remove code barriers, create incentives, or fill regulatory gaps. Each recommendation has a corresponding "brief," designed by and for public officials and staff across the country. The briefs consist of:

- *Introduction*–explaining the recommended ordinance to amend the code;

- *Effects*–detailing how adopting the recommended ordinance may affect the community, including costs and benefits; and

- *Examples*–describing in plain language 2-4 examples of enacted ordinances adopting the recommendation; this section also includes links and citations for 4-6 additional ordinances adopted by local governments.

When fully drafted, the SDC will consist of over 1,500 recommendations. After the SDC is fully drafted, one-third of the chapters will be updated every year.

Endnotes

1. ITALO CALVINO, INVISIBLE CITIES 51 (1972).
2. *See* Hyuna Sung, et al., *Emerging cancer trends among young adults in the USA: analysis of a population-based cancer registry,* THE LANCET, Feb. 3, 2019 (study indicating increases in obesity are leading to increases in obesity-related cancer).
3. *See* Gabriel Zucman, *Global Wealth Inequality,* NAT. BUREAU OF ECONOMIC RESEARCH, Jan. 2019, https://perma.cc/7F78-Q6FP (noting that income inequality increased since the 1980s and is now greater than any year after the Great Depression).
4. *See* Erin Ryan, *Memo to Environmentalists: Brace for Preemption, Propertization, and Problems of Political Scale in* ENVIRONMENTAL LAW. DISRUPTED., (ELI Press forthcoming 2020) ("With federal environmental law under sustained attack since 2016, it becomes incumbent on environmental advocates to think more seriously about how to continue pursuing solutions to national-level environmental problems by means other than federal authority."); Victor B. Flatt & Rob Verchick, *Burying Our Head in Sand on Climate Change No Longer an Option,* HOUS. CHRON. (Sept. 28, 2017), http://www.houstonchronicle.com/opinion/outlook/article/ Burying-our-head-in-sand-on-climate-change-no-12238961.php.
5. Susan Crawford, *Autonomous Vehicles Might Drive Cities to Financial Ruin,* WIRED.COM, June 20, 2018.
6. Intergovernmental Panel on Climate Change, *Global Warming of 1.5 C* (2018), https://perma.cc/2UBR-SAPZ; U.S. Global Change Research Program, *Fourth National Climate Assessment* (2018), https://perma.cc/7GFG-PE9V.
7. IPBES Global Assessment, *Report of the Plenary of the Intergovernmental Science-Policy Platform on Biodiversity and Ecosystem Services* (May 2019), https://www.ipbes.net/global-assessment-report-biodiversity-ecosystem-services.
8. Sandra L. Colby & Jennifer M. Ortman, *Projections of the Size and Composition of the U.S. Population: 2014 to 2060* (U.S. Census Bureau 2015) (projecting 2040 U.S. population to be almost 400 million). As of July 2019, U.S. population was approximately 330 million. *See U.S. and World Population Clock,* https://www.census.gov/popclock/.
9. ARTHUR NELSON, PLANNER'S ESTIMATING GUIDE: PROJECTING LAND-USE AND FACILITY NEEDS (Routledge 2004).
10. *Id.*
11. *About,* Sustainable Development Code, http://sustainablecitycode.org/about/ (last visited May 25, 2019).
12. *Id.*

Introduction

This book is the first to index and reprint a subchapter of the SDC. Due to the immediacy of the climate change challenge, we chose to print the Climate Change Subchapter in full first to make it as accessible as possible. Our hope is that it will be useful to local communities regardless of their population and location and that it will help expedite the mitigation of greenhouse gas emissions (GHGs). As the SDC rolls out future chapters we hope to make them available in print format.

This book provides local governments with a diversity of approaches to reduce GHGs and/or increase natural features that absorb GHGs, such as trees and wetlands. This book focuses on actions that are traditionally within local governments' land use and development authority. The recommendations focus exclusively on enacted ordinances (i.e. not policies or informal statements) that are part of the development code. While some SDC recommendations may overlap with building codes and comprehensive plans, we are as cognizant as possible to keep the recommendations within the development code and process.

Although SDC recommendations cite and describe enacted ordinances, each community should ensure that newly enacted ordinances are within their specific local authority, have not been preempted, and are consistent with state comprehensive planning laws. Also, the effects described in SDC recommendations are based on existing examples. Those effects may or may not be replicated elsewhere. If you adopt an ordinance or adopt a GHG reduction ordinance not part of the SDC, we would very much like to hear about your experience.

Finally, we believe that the climate change recommendations in this book are critical to helping localities and states meet their GHG reduction targets. If your community would like to discuss any of the recommendations, please contact us at: https://sustainablecitycode.org/share/.

Part 1:

REMOVE CODE BARRIERS

ACCESSORY DWELLING UNITS

Tyler Adams (author)
Jonathan Rosenbloom & Christopher Duerksen (editors)

INTRODUCTION

Accessory dwelling units (ADUs) help maximize space in residential districts. ADUs are additional living quarters on single-family lots that are independent of the primary dwelling unit and can go by a variety of names such as accessory apartments, second units, in-law units, laneway houses, and granny flats.[1] These additional apartments or cottages may or may not be attached to the existing home and tend to be smaller, with many jurisdictions having size limitations. ADUs are built with fully-equipped living quarters, including kitchens and baths. ADUs are typically owned by the property owner of the single-family lot and are either rented out or granted access to another party depending on the situation and permission pursuant to the local code.[2] While ADUs can take a variety of forms, they generally appear in three configurations in relation to the primary dwelling: internal, attached, and detached. Internal ADUs are integrated into the existing structure, such as a converted attic or basement, attached ADUs are built on as additions to the primary unit, and detached ADUs are built structurally separate, such as a converted detached garage.[3]

ADUs are not a new concept to address affordable housing. In the United States, ADUs were a very popular housing option in the early nineteenth century with the expansion of cities during industrialization.[4] However, following World War II the American shift to suburban-ism led to many leaving the city and seeking out larger single-family lots that were restricted by Euclidean zoning.[5] The rise in housing prices and the decrease in the availability of low-rent units has led many jurisdictions to bring back ADUs as a form of low impact affordable housing. Even outside the legislative process communities have been developing programs using ADUs as a way to combat housing shortages and the subsequent rise in homelessness.[6]

EFFECTS

ADUs respond to the desire to have more small, discreet, affordable, traditional, and environmentally friendly forms of housing.[7] Many people do not require or wish to have the type of large homes that are often mandated in single family districts. Yet, many people also want to enjoy the amenities that accompany these neighborhoods. ADUs not only provide an opportunity to receive rental income and offer multigenerational housing options, but also help maximize space.[8] Further, they do so without changing the essential characteristics of a neighborhood, allowing residents to still retain the benefits of living in the community.[9]

Perhaps most importantly, ADUs help increase a community's housing supply and offer affordable housing options to many low- and moderate-income residents.[10] ADUs are a way to create affordable housing at low cost because there is no need to develop new infrastructure due to their integration into existing facilities and utilities.[11] ADUs offer an alternative that helps create new housing in low density neighborhoods, while keeping developer and municipality costs comparatively lower.

As a type of infill development, ADUs help increase neighborhood densities to levels where other modes of transit become viable alternatives to automobiles, thus reducing the amount of greenhouse gases being emitted.[12] Also, because ADUs are typically smaller than the average single-family home fewer natural resources are consumed in the construction process as well as the during the day-to-day operation of the unit.[13]

EXAMPLES

Ann Arbor, MI

In 2016, Ann Arbor, MI approved zoning amendments that greatly relaxed the requisite standards for ADUs. The City Council proposed the amendments because of the 2015 Housing Affordability and Economic Equity Analysis which documented increasing housing costs in Ann Arbor.[14] As a result, moderate income and working families as well as young adults and seniors were being priced out of the housing market.[15] Previously, ADUs had to be designed so that the appearance of the building remained that of a single-family residence. The amendment altered these requirements such that ADUs can now be designed to match either a single-family residence or a detached accessory building, such as a garage or carriage house.[16] Perhaps the most dramatic change is that of the occupancy requirements. The old ordinance required that ADUs only be occupied by relatives or a maximum of two employees of the homeowner and no rent could be charged.[17] The amendment states that the property owner is required to occupy either the ADU or the main residence, removing the previous restriction.[18] It further prohibits leasing or renting of the ADU for less than 30 days, thus allowing the homeowner to lease to anyone long term.[19]

To view the provision see Ann Arbor, MI Code of Ordinances § 5:10.2 4(d) (2016).

Barnstable, MA

The town of Barnstable, recognizing the essential role accessory apartments were playing in filling the demand for afford-

able housing, enacted a series of ordinances designed to bring into compliance previously unpermitted accessory apartments as well as providing for the creation of new ones.[20] The amnesty program provides that accessory dwelling units that violate the zoning code ordinances are able to come in compliance provided that they comply with certain criteria. One of these criteria being that upon receipt of a comprehensive permit, the owner of the accessory apartment agrees to rent to a person or family whose income is 80% or less of the area median income.[21] In addition, the rent to be charged cannot exceed the rents established by the Department of Housing and Urban Development for a household whose income is 80% or less of the median income for the area.[22] In order to eligible for the construction of a new accessory apartment, a homeowner must also comply with the previous restrictions.[23] There is also a deed restriction that homeowners of previous and new accessory apartments must execute which will restrict any future rentals by future owners to persons whose income is 80% or less of the median income.[24]

To view the provision see Town of Barnstable, MA Code § 9-12 (2002).

ADDITIONAL RESOURCES

Sage Computing, Inc., *Accessory Dwelling Units: Case Study*, HUD User (June 2008), https://perma.cc/JLQ6-NEVQ.

Martin J. Brown, *Accessory Dwellings: What They Are and Why People Build Them*, Accessory Dwellings, https://perma.cc/W4FC-4HUC, (last visited April 19, 2018).

Jonathan Coppage, *Accessory Dwelling Units: A Flexible Free-Market Housing Solution*, R Street (Mar. 2017), https://perma.cc/CDZ6-L7ZQ.

ENDNOTES

1   Sage Computing, Inc., *Accessory Dwelling Units: Case Study*, HUD User (June 2008), https://perma.cc/JLQ6-NEVQ.
2   *Id.*
3   Tran Dinh et al., *Yes, In My Backyard: Building ADUs to Address Homelessness*, University of Denver Sturm College of Law Homeless Advocacy Policy Project (May 3, 2018), available at SSRN: https://perma.cc/5NWQ-4FCR.
4   Sage Computing, Inc., *supra* note 1.
5   *Id.*
6   Dinh et al., *supra* note 3. Seattle's BLOCK Project is a crowdfunded program that places pre-fabricated ADUs on participating homeowner's lots. Unhoused individuals are then matched with homeowners and are able to reside in the ADU.
7   Martin J. Brown, *Accessory Dwellings: What They Are and Why People Build Them*, Accessory Dwellings, https://perma.cc/W4FC-4HUC (last visited April 19, 2018).
8   *Id.*
9   Jonathan Coppage, *Accessory Dwelling Units: A Flexible Free-Market Housing Solution*, R Street (Mar. 2017), https://perma.cc/CDZ6-L7ZQ.
10   *Id.*
11   Sage Computing, Inc., *supra* note 1.
12   *Small Backyard Homes: Accessory Dwelling Units (ADUs)*, Portland State University, https://perma.cc/W9S7-49PW (last visited May 22, 2018).
13   *Id.*
14   City of Ann Arbor, *Accessory Dwelling Units*, Planning, https://perma.cc/9HEN-AXBR (last visited May 21, 2018).
15   *Id.*
16   City of Ann Arbor, *Approved Amendments to the Zoning Ordinance (8-4-2017)*, https://perma.cc/NL44-ASG8 (last visited May 17, 2018).
17   *Id.*
18   *Id.*
19   *Id.*

20 Town of Barnstable, MA Code § 9-12 (2002).
21 *Id.* § 9-14.
22 *Id.*
23 *Id.* § 9-15.
24 *Id.*

CLUSTER/CONSERVATION SUBDIVISION IN RURAL/ URBAN AREA

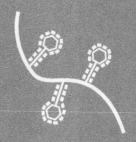

Alec LeSher (author)
Jonathan Rosenbloom & Christopher Duerksen (editors)

INTRODUCTION

Clustered/Conservation Subdivision (CCS) ordinances allow residential developments in rural and urban fringe areas while retaining the natural character and significant wildlife habitat in the newly developed area. CCS developments are an alternative to traditional residential developments (often subdivisions and/or planned unit developments) that typically result in substantial destruction of natural features and habitat.[1] In contrast, CCS ordinances allow or require dense clusters of residential units in one part of the proposed project area, in exchange for permanently preserving open space and natural features.[2] Under this model, the open space is rendered undevelopable, but may be used by the community for recreation, agriculture, or conservation (for alternative ways to zone the PUD open space see Limit PUDs Near Sensitive Natural Areas). Placing CCS developments near the urban/rural boundary helps provide a natural transition from dense urban areas to more open and natural spaces.

Local governments may structure CCS ordinances in a variety of ways. Some local governments have chosen to create overlay districts that indicate where a CCS may be located.[3] Others may choose to allow a CCS directly in the regulations for various zones, most typically residential zones.[4] The ordinance should also describe the design standards for the CCS, such as maximum allowable surface coverage, minimum amount of open space required, and where or how the new residences will be clustered on the parcel. Local governments should note that CCS developments often require smaller minimum lot sizes than the underlying zoning district would otherwise allow to accommodate for tight clusters of buildings, so that more open space may be preserved. Many jurisdictions simply exempt CCS developments from minimum lot size requirements.[5] In many ordinances, there is a formula or ratio that indicates how much space developers must leave open. These formulas and ratios may be based on prior use patterns for new developments in communities that have traditionally had access to public open spaces, or they can be tailored to meet the needs of developments in communities that have historically been without open spaces. Finally, it is important for a municipality to specify what happens to the development rights of the preserved open space. Some local governments retain those development rights indefinitely, while others allow the transfer of those rights to non-profit land trust entities or a homeowners association.[6]

EFFECTS

Promoting CCS developments provides numerous benefits to local governments. First, preserving open space and wildlife habitat provides a natural corridor for wildlife to travel, and therefore increases the biodiversity in the jurisdiction.[7] One study found a strong correlation between using cluster type developments and the increased preservation of wildlife habitat and biodiversity, as compared to traditional residential developments, which had less wildlife habitat and biodiversity.[8] Second, the preservation of open space allows existing greenspace to continue to provide ecosystem services. Greenspace is open, undeveloped, or vegetated land that captures greenhouse gas (GHG) emissions and allows wildlife to move freely through a natural landscape.[9] The services that greenspace provide also include air purification, stormwater management and treatment, and soil retention.[10] Finally, one study found that homes in CCS developments gain monetary value at a quicker rate than traditional residential subdivisions.[11] Both homeowners and local governments may benefit from the resulting increase in assessed property value.

In contrast, a traditional residential subdivision consumes almost all of the greenspace on a parcel, which shifts costs to local governments to replace the loss of ecosystem services. Traditional residential subdivisions can also damage or destroy wildlife habitat and block existing wildlife corridors, thereby pushing wildlife out of the jurisdiction and harming overall biodiversity.[12] CCS ordinances mitigate these harms by allowing residential developments that complement the natural environment and preserve substantial wildlife corridors and habitat.[13]

Thurston County, WA

Thurston County is home to the state capitol of Olympia, which sits near the northern edge of the county border. Olympia has a large metropolitan area, but much of the County is rural and zoned as "Long Term Agriculture" (LTA) or "Long Term Forestry" (LTF).[14] Within these zones the County requires low density housing, which in some instances can mean one dwelling unit per eighty acres.[15] However, a landowner in a LTF or LTA district can apply to create a "Planned Rural Residential Development" (PRRD).[16] These PRRDs are exempt from minimum lot size requirements, allowing a developer to maximize profits while also retaining the rural character of the area.[17]

PRRDs require the creation of a "resource parcel."[18] The composition of the resource parcel varies based on the underlying zoning district. For instance, in LTA districts the parcel must be used for agriculture, while in LTF districts the resource parcel must consist of forested land.[19] The percentage of the lot dedicated to the resource parcel also varies. In LTA districts, eighty-five percent of the PRRD must be a resource parcel.[20] In LTF districts, seventy five percent of the PRRD must be a resource parcel.[21] In all cases, the resource parcel must be one contiguous area of land, not intermingled with any residential developments.[22] This requirement insures that the residential development is limited to a small portion of the total platted area, thereby preserving natural wildlife corridors and habitat.

The County further regulates how residences are situated in the PRRD. Homes must blend in with the natural features as seen from the public roadway, and the configuration and size of lots must vary.[23] Lots must be grouped, rather than arranged in a linear fashion.[24] The goal of these requirements is to provide unhindered access and use of the resource parcel, and promote a pleasant rural aesthetic as viewed from the highway.[25]

To view the provisions see Thurston Cty., WA, Code of Ordinances § 20.30A (1993).

Jamestown, RI

Jamestown is located on an island just West of Newport, Rhode Island. A majority of the jurisdiction is zoned for low-density residential uses.[26] Within three of the largest of those zones, clustered development is *required* for any subdivision of land over five acres (emphasis added).[27] The planning commission will only approve a traditional type residential development if it determines that a clustered development is inappropriate due to "land configuration, prevailing development adjacent to the parcel, or environmental condition."[28]

In a cluster development, the town requires that fifty percent of the area be devoted to open space[29] (see Limit PUDs Near Sensitive Natural Areas). Permitted uses in the open space are limited to conservation, recreation, agriculture, and preservation of historic sites.[30] However, a developer may construct certain structures in the open space, such as walkways, retaining walls, recreational facilities, or utilities.[31] Notably,

any open space that has been deemed "unsuitable for development" will not count towards the fifty percent requirement.[32] The open space must then be conveyed to the Town, to a cooperative or homeowners association, or donated to a non-profit land trust entity.[33]

The Town also sets minimum lot sizes for clusters based on the underlying zoning district. For example, in the zone designated as rural-residential that normally requires minimum lot sizes of 200,000 square feet, that requirement is reduced to 20,000 square feet.[34] On the other end of the spectrum, the zone that normally requires at least 40,000 square feet lots only requires 8,000 square feet lots in cluster developments.[35]

To view the provisions see Jamestown, RI, Code of Ordinances §§ 82-1600 to 1608 (2003).

Baltimore County, MD

Baltimore County is a largely rural jurisdiction with the City of Baltimore on its Southern border. The County's zoning regulations establish several zones designed to preserve rural resources and the natural character of the county.[36] CCS developments are required in some cases for the subdivision of land in certain "Resource Conservation" districts. [37]

For example, in the Watershed Protection zone, any development of a parcel of land greater than ten acres must be a clustered design.[38] 70% of the tract must be reserved as the "conservancy area."[39] Whenever possible, the conservancy are must be a contiguous block of land that contains valuable ecological features such as prime soil, steep slopes, wetlands, and forests.[40] The conservancy area is required to be held by a single entity, such as a land trust or homeowner's association, and a permanent preservation easement must be placed over the area.[41] The entity that owns the area must also file an agreement with the county to take responsibility for maintaining the area and preserving it as is.[42]

To view the provisions see Baltimore Cty., MD, Zoning Regulations § 1A03.5 (1992).

ADDITIONAL RESOURCES

Guidelines for Calculation and Provision of Open Space in Developments of Regional Impact Technical Bulletin 94-001, as Amended, Cape Cod Commission 5-7 (May 4, 2009).

Public Dedication of Land and Fees-in-Lieu for Parks and Recreation: A Tool for Meeting Recreational Demands in Pennsylvania Municipalities, Pennsylvania Land Trust Association (2015).

ENDNOTES

1   Charlotte E. Gonzales-Abraham et al., *Patterns of Houses and Habitat Loss from 1937 to 1999 in Northern Wisconsin, USA*, 17 ECOLOGICAL APPLICATIONS, 2011, 2017 (2007).
2   RANDALL ARENDT ET AL., RURAL BY DESIGN: MAINTAINING SMALL TOWN CHARACTER 229-32 (1994).
3   *See, e.g.*, JAMESTOWN, R.I., CODE OF ORDINANCES § 82-1600 (2003).
4   *See, e.g.*, GRAYSLAKE, ILL., ZONING ORDINANCE § 17.32.100 (D) (current through 2018).
5   *See, e.g.*, LOUDON CTY., VA., CODE OF ORDINANCES § 1226.02 (a) (1998); Concord, MA, Zoning Bylaw § 9.1-9.5 (1962).
6   *See, e.g.*, JAMESTOWN, R.I., CODE OF ORDINANCES § 82-1608 (2003).
7   John Roach, *First Evidence that Wildlife Corridors Boost Biodiversity, Study Says*, NAT'L GEOGRAPHIC NEWS, Sept. 1, 2006, at https://perma.cc/RE8J-2LMT.
8   Gonzales-Abraham, *supra* note 1, at 2017.

9   U.S. Envt'l Prot. Agency, *What is Open Space/Green Space?*, https://perma.cc/ET63-53V6 (last visited May 18, 2018).

10 J.B. Ruhl, *The Twentieth Annual Lloyd K. Garrison Lecture: In Defense of Ecosystem Services*, 32 PACE ENVTL. L. REV. 306, at 309 (2015).

11 JEFF LACY, AN EXAMINATION OF MARKET APPRECIATION FOR CLUSTERED HOUSING WITH PERMANENTLY PROTECTED OPEN SPACE 1-12 (1990).

12 Stephen DeStefano & Richard M. DeGraff, *Exploring the Ecology of Suburban Wildlife*, 1 FRONTIERS IN ECOLOGY & THE ENV'T, 95, 101 (2003).

13 *See* ARENDT ET AL., *supra* note 2.

14 *Official Zoning Map of Thurston County, Washington*, Thurston Cty., Wash., (July 15, 2009), https://perma.cc/EX4E-JRM8.

15 THURSTON CTY., WASH., CODE OF ORDINANCES § 20.08D.045 (2012).

16 *Id.* § 20.30A (1993).

17 *Id.* § 20.30A.070.

18 *Id.* § 20.30A.040 (1).

19 *Id.* § 10.30A.040 (3).

20 *Id.* § 20.30A.040 (1).

21 *Id.*

22 *Id.* § 20.30A.070 (5) (b).

23 *Id.* § 20.30A.070 (6) (a).

24 *Id.* § 20.30A.070 (6) (d).

25 *Id.* § 20.30A.070 (6) (d) (explanatory note).

26 *Jamestown Zoning Map*, Town of Jamestown, Rhode Island (2009), https://perma.cc/7A5Z-WTK9.

27 JAMESTOWN, R.I., CODE OF ORDINANCES § 82-1602 (2003).

28 *Id.*

29 *Id.* § 82-1604.

30 *Id.* § 82-1606.

31 *Id.* § 82-1607.

32 *Id.* § 82-1606.

33 *Id.* § 82-1608.

34 *Id.* §§ 82-1604, 82-800.

35 *Id.* §§ 82-1604, 82-302.

36 *See* BALTIMORE, MD., ZONING REGULATIONS § 100.1 (A) (2) (1975).

37 *See, e.g., id.* § 1A03.4 (B) (1) (b).

38 *Id.*

39 *Id.* § 1A03.4 (B) (1) (b) (1).

40 *Id.* § 1A03.5 (A) (1) (a)-(g).

41 *Id.* § 1A03.5 (C) (1)-(2).

42 *Id.*

DISTRICT HEATING AND COOLING ZONES

ZONE

Kyler Massner (author)

Jonathan Rosenbloom & Christopher Duerksen (editors)

INTRODUCTION

District heating and cooling systems, commonly called district energy systems (DES), provide heating and cooling to buildings that are connected to and powered by localized utility plants.[1] DES can meet the energy needs of a variety of community sizes in a cost efficient, renewable, and reliable manner.[2] DES is best suited in zones with a mix of medium to high density developments, such as college or government campuses, downtown districts, airports, mixed use residential clusters, industrial parks, and healthcare facilities.[3] DESs have been used since the late 1800's, but local governments have increasingly turned to them to respond to volatile energy prices, the need for energy efficiency, concern for greenhouse gas (GHG) emissions, and a desire to increase resilience by decentralizing infrastructure and utilities.[4]

Despite DESs' long history, barriers to their construction exist in many local development codes. Typically, community energy needs are considered late in the planning process and envision the use of a centralized energy utility. Local zoning ordinances may prohibit smaller scale generators from being constructed in particular zones explicitly or, more commonly, implicitly by not including them in permitted uses.[5] Local governments can remove these barriers by addressing energy infrastructure earlier in the development process to ensure that future developments are coordinated to leverage the benefits of DESs.[6] Additionally, local governments can expand opportunities for installation of small scale DESs by amending zoning ordinances to permit construction in residential and commercial zones.[7]

DESs typically have three core components: the thermal energy generating powerplant, the distribution system, and the energy transfer station.[8] The generating powerplant can be a traditional powerplant reliant on fossil fuels, a renewable energy system, or a combined heat and power system (CHP). The product produced is either steam, hot water, or chilled water, that is then pumped through a distribution system of heavily insulated pipes, providing heat, cooling, and/or hot water to buildings that are connected via a series of energy transfer stations (e.g., meters, valves, pumps).[9] The end product is distributed to connected buildings via energy transfer stations.[10] The three parts of a DES typically create a closed loop energy system (i.e. energy provided to the building is used and then returned to the powerplant for reuse), thus creating highly efficient systems that eliminate waste experienced with traditional energy systems.[11]

EFFECTS

DESs provide localized energy from diverse fuel sources, while building community resilience to future energy challenges.[12] Decentralizing power producing facilities to place them closer to populated areas reduces energy loss that escapes during the longer distribution distances typical of centralized utilities.[13] Additional benefits stemming from DESs include increased security from the volatile traditional energy market, increased system reliability, reduced maintenance and energy costs, and reduced GHG emissions.[14] The implementation of DESs also eliminates the need to house individual heating or cooling systems on the property. This frees space for other uses, including conservation, new projects, or improvements.[15]

DESs can stabilize and reduce energy costs by producing and storing thermal energy from sustainable resources at times of low demand.[16] Energy efficiency is increased by the elimination of distribution losses due to a closed loop energy system.[17] Because energy efficiency is improved, DESs can lead to lower utility bills, allowing a community to be economically competitive and attractive to start-up businesses and developers due to a reduction in operating costs.[18] In addition, reductions in utility bills make the district more affordable to low-income individuals and groups.[19]

DESs have flexibility in the type of fuel needed to operate the system. Relying on diverse fuel types insulates users from market fluctuations in a single energy source.[20] For example, as described below, St. Paul, Minnesota created a DES that integrates natural gas, fuel oil, CHP, and solar energy to replace a tradition coal fueled power plant. In doing so, St. Paul is able to stabilize pricing.[21]

EXAMPLES

St. Paul, MN

The City of St. Paul, working with the State of Minnesota, the U.S. Dep't of Energy, and the local downtown business community, launched District Energy St. Paul in 1983 to build and operate a DES.[22] Codified in Section 405 of the St. Paul Code of Ordinances, the City granted a franchise to District Energy St. Paul, Inc. to operate a DES within St. Paul and sell the energy to residents of the City.[23] In return, the City receives fees from District Energy and regulates District Energy's rates.[24]

District Energy St. Paul is considered one of the most notable DESs in part because its CHP system utilizes a variety of different fuels.[25] The ability to have flexible fuel options allows District Energy to maximize biomass utilization, save money, and reduce GHG emissions. For example, District Energy relies on recycled wood chips to provide the primary fuel for the CHP.[26] These wood chips are the product of urban wood residuals, such as tree trimmings and leftover construction material, that are delivered from sites within 60 miles of the plant.[27] Utilizing approximately 280,000 tons of urban wood residuals, District Energy's CHP is able to produce 65 megawatts of heat and up to 33 megawatts of electricity that provides enough power for 20,000 homes.[28] Taking advantage of a renewable resource provides a stable energy supply, improved energy security, and reduced GHG emissions.[29] In October 28, 2015, District Energy announced that it had begun plans to eliminate the use of coal by 2021, a measure that would reduce CO_2 emissions by 27%, the equivalent of removing 4,400 cars from the road.[30]

District Energy St. Paul was also the first in the U.S. to integrate solar thermal into the DES. By installing a 23,000 square foot system of 144 thermal solar collectors, primarily used for the St. Paul RiverCentre, the system can generate 1,000 MWh of energy each year and can export any extra energy produced to the grid.[31]

To view the provision see Saint Paul, MN, Code of Ordinances, App. F §§ 1,2, 5-6 (2007).

Schaumburg, IL

Schaumburg, IL explicitly encourages the safe, effective, and efficient development of DESs in its zoning code, permitting both commercial and residential construction of DESs in all village zoning districts. The only restriction which the City imposes is that the appearance be constructed with similar characteristics to the surrounding buildings. Schaumburg also reduces barriers by providing clear guidelines for construction of DES, allowing the network of conduit piping to be placed either within easements on a lot or within a vehicular right-of-way. Clear and specific development ordinances benefit developers by reducing ambiguity surrounding construction and are an example of how a preplanned and coordinated energy plan can lead to successful implementation.

To view the provision see Schaumburg, IL, Code of Ordinances § 154.70 (C) (2018).

Detroit, MI

Detroit Thermal has reliably provided steam for heating, hot water, and absorption chilling to 30 million square feet of downtown Detroit for more than 100 years.[32] Detroit Thermal relies on Detroit Renewable Power as the primary source for its steam. DRP's facility provides clean, efficient, and reliable steam by converting up to 3,300 tons of municipal solid waste per day into fuel to power large steam producing commercial boilers.[33] DRP's waste-to-energy plant generates 68 megawatts of electricity with the ability to export up to 550,000 pounds per hour of steam.[34] Detroit's system of renewable resource integration allows Detroit to reduces its waste volume by 90% and deliver thermal service from a clean, renewable energy source which reduces Detroit's carbon footprint.

To view the provision see Detroit, MI, Code of Ordinances § 6-509 (2018).

ADDITIONAL RESOURCES

Informing, Connecting & Advancing the District Energy Industry, International District Energy Association, https://perma.cc/6MFC-G8RL (last visited June 13, 2018) (homepage for the International District Energy Association).

Combined Heat and Power Partnership, *Catalog of CHP Technologies*, EPA (Sept. 2017), https://perma.cc/A63W-HKA5 (last visited May 30, 2018) (containing an overview and a cost/benefit analysis of popular CHP technologies from the EPA).

Office of Sustainable Communities, *District-Scale Energy Planning: Smart Growth Implementation Assistance to the City of San Francisco,* EPA (June 2015) https://perma.cc/9HU5-AQ2F (last visited May 24, 2018) (providing a four-step implementation guide for district energy planning).

Lauren T. Cooper & Nicholas B. Rajkovich, *An Evaluation of District Energy Systems in North America: Lessons Learned from Four Heating Dominated Cities in the U.S. and Canada,* Lawrence Berkeley National Laboratory (Aug. 2012), https://perma.cc/97QY-7KU4 (last visited May 24, 2018) (North American district energy system case studies).

CREATIVE ENERGY VANCOUVER PLATFORMS INC., CREATIVE ENERGY NES: BUILDING COMPATIBILITY DESIGN GUIDE 6-7 (Oct. 2015), https://perma.cc/MDF2-FD4S (last visited May 30, 2018) (summarizing building design strategies for developers).

ENDNOTES

1   NATIONAL RESOURCE COUNCIL, DISTRICT HEATING AND COOLING IN THE UNITED STATES: PROSPECTS AND ISSUES 7 (National Academies Press 1985).
2   U.S. ENERGY INFORMATION ADMINISTRATION, U.S. DISTRICT ENERGY SERVICES MARKET CHARACTERIZATION 4 (DOE Feb. 2018), https://perma.cc/HU4V-ZFH4 (last visited May 30, 2018).
3   *Id.* at 16; LAUREN T. COOPER & NICHOLAS B. RAJKOVICH, AN EVALUATION OF DISTRICT ENERGY SYSTEMS IN NORTH AMERICA: LESSONS LEARNED FROM FOUR HEATING DOMINATED CITIES IN THE U.S. AND CANADA 1-2 (Lawrence Berkeley Nat' Lab. Aug. 2012), https://perma.cc/97QY-7KU4.
4   U.S. ENERGY INFORMATION ADMINISTRATION, *supra* note 2 at 19; COOPER & RAJKOVICH, *supra* note 3, at 1-2.
5   MARLENA ROGOWSKA, DISTRICT ENERGY WITHIN THE PLANNING CONTEXT: EXPLORING THE BARRIERS AND OPPORTUNITIES FOR DISTRICT ENERGY AND COMMUNITY ENERGY SOLUTION IN ONTARIO, CANADA 8-10 (Ryerson University 2013), http://perma.cc/HKT4-NAXF (last visited June 5, 2018).
6   *Id.*
7   *See id.*

8   Steve Tredinnick, *Why is District Energy Not More Prevalent in the U.S.?*, HPAC Engineering (June 7, 2013), https://perma.cc/97PW-3PDZ (last visited May 25, 2018).

9   U.S. ENERGY INFORMATION ADMINISTRATION, *supra* note 2, at 4.

10   CREATIVE ENERGY VANCOUVER PLATFORMS INC., CREATIVE ENERGY NES: BUILDING COMPATIBILITY DESIGN GUIDE, 6-7 (Oct. 2015), https://perma.cc/MDF2-FD4S (last visited May 30, 2018).

11   COOPER & RAJKOVICH, *supra* note 3, at 2.

12   ROGOWSKA, *supra* note 5, at 6-7.

13   *Id.* at 8.

14   U.S. EPA, OFFICE OF SUSTAINABLE COMMUNITIES, DISTRICT-SCALE ENERGY PLANNING: SMART GROWTH IMPLEMENTATION ASSISTANCE TO THE CITY OF SAN FRANCISCO 6-7 (June 2015) https://perma.cc/9HU5-AQ2F (last visited May 24, 2018).

15   *Id.*

16   *Id.*

17   *Id.*

18   *Id.*

19   *Id.*

20   *Id.*

21   *Id.*; Ken Smith, *The Wave - Spring 2014*, District Energy St. Paul (Apr. 2018), https://perma.cc/F7KR-77WU (last visited May 25, 2018).

22   NATIONAL RESOURCE COUNCIL, *supra* note 1 at 13; *Combined Heat and Power*, District Energy St. Paul, https://perma.cc/U3S5-XERT (last visited May 25, 2018); *History*, District Energy St. Paul, https://perma.cc/BHJ4-TWM6 (last visited May 25, 2018).

23   SAINT PAUL, MINN., CODE OF ORDINANCES, App. F §§ 1, 2 (2007).

24   *Id.* § 5-6.

25   *History, supra* note 22.

26   *Combined Heat and Power, supra* note 22.

27   *Id.*

28   *Id.*

29   *Id.*

30   *District Energy St. Paul Plans Elimination of Coal for Heating System*, District Energy St. Paul (Oct. 28, 2015), https://perma.cc/T88G-QE6Q (last visited May 25, 2018).

31   U.S. ENERGY INFORMATION ADMINISTRATION, *supra* note 2, at 54.

32   *Welcome to Detroit Thermal*, Detroit Thermal, https://perma.cc/3R8K-7Y49 (last visited May 30, 2018).

33   *Steam Energy from Renewable Sources*, Detroit Renewable Power, https://perma.cc/3HWH-KS8X (last visited May 29, 2018).

34   *Id.*

HEIGHT & SETBACKS TO ENCOURAGE RENEWABLES

Kerrigan Owens (author)
Jonathan Rosenbloom & Christopher Duerksen (editors)

INTRODUCTION

Height and set back requirements can frustrate the use and installation of wind and solar power systems in urban areas. Because solar and wind systems are often installed on rooftops, they are often considered part of the structure and calculated towards a buildings' maximum height. As such, developers building near the maximum allowable height may not be permitted to install rooftop energy systems. If a developer wanted to install a renewable system on the roof, she would be required to reduce the overall height of the interior space to make room for the solar or wind system or seek a variance. This may reduce the square footage or make development more expensive, making it more difficult for the developer to meet their financial expectations. To encourage more developers to incorporate renewable energies this ordinance would relax the height and setback requirements in relation to wind and solar energy systems.[1] This ordinance can be drafted in a way to reduce requirements across districts or to create specific exceptions to height or setbacks.

Another tool being used by local governments is the incorporation of the International Building Code (IBC).[2] The IRC is updated every three years and includes the best practices from around the nation.[3] One of the recent additions to the IRC is the "Solar-Ready Provision" which details how to expedite and increase smaller scale solar units on homes. Some of these enhancements include constructing homes with minimal rooftop equipment, orienting buildings in a "north-south" fashion and providing a detailed plan of the roof so that solar installations can confirm that roof will be able to support the systems.[4] Several local governments have also begun to require on-site renewable energy capacity prior to issuing a certificate of occupancy (for more information see Zero Net Energy Buildings brief).[5]

To further promote renewable energy systems, several local governments permit solar and wind systems by-right (for more information see the brief Allow Solar Energy Systems and Wind Turbines by-Right). In addition, Oregon has enacted statewide legislation known as the "Oregon Solar Installation Specialty Code," which establishes setback and height requirements that pre-empt city codes.[6]

EFFECTS

Electrical energy is one of the largest demands of fossil fuels.[7] Burning these fuels releases carbon dioxide and other pollutants into the atmosphere which increases greenhouse gas (GHG) emissions, leading to climate change.[8] One way to lower GHG emissions is to switch to alternative energy production.[9] Relaxing local regulatory requirements around alternative energy production will encourage more individuals to choose alternative energies on their own accord without direct government actions. These alternatives help to decrease air pollution and help improve human health[10] by mitigating respiratory illnesses that can stem from the burning of fossil fuels.[11] By relaxing regulations on solar and wind energy systems, citizens are able to choose systems that help mitigate the effects of climate change while improving and promoting human health.[12]

EXAMPLES

Minneapolis, MN

Minneapolis like most municipalities has height and setback requirements in each of its zoning districts. These requirements would in many cases frustrate the construction of wind and solar power systems. However, Minneapolis codified separate ordinances to govern wind and solar energy systems in all districts.[13] One ordinance sets universal standards for wind production in all districts. This ordinance specifically governs wind systems while solar systems are governed by other sections of the code. Wind energy systems are limited to a height of 15 feet measured from where the turbine is attached either to the building or to the ground.[14] These attached systems are permitted by right.[15] On buildings that are over four stories, the wind energy system must be installed above the fourth story.[16] Free standing systems are permitted on a conditional basis, and must comply with condition specific standards such as encroachments and setbacks, specific height requirements per zoning district, etc.[17] The provision also lists specific aesthetic requirements that the wind systems must meet, including using compatible materials, colors, and textures of surrounding buildings.[18]

The ordinance for solar systems functions in a similar fashion to the wind system ordinance and allows solar systems by right within all zoning districts.[19] The solar ordinance sets height requirements for solar systems to not extend further than three feet above the ridge level roof and cannot extend further than ten feet above surface roof. The setback requirement for solar systems is one foot from the perimeter of the roof, but for any system

which does not extend above three feet there is no setback requirement.[20] For freestanding solar systems, they must be constructed to stand below twenty feet or to not exceed the principal structure.[21] The ordinance also sets a requirement that for solar systems within a residential or office district, the system may not exceed five percent of the total lot area.[22] Furthermore, the ordinance states that even if the solar system does not meet the above criteria, there is the option of applying for a permit for conditional use.[23] This application mirrors the above criteria, and allows for an administrative evaluation of the specific applicant.[24]

To view the wind energy provision see Minneapolis, MN, Zoning Code §§ 535.710, 535.840 (2007).

Laramie, WY

Laramie's energy code allows for both solar and wind energy systems in all zoning districts, with no exceptions for homeowners association restrictions.[25] Solar energy systems are given an additional three feet of space above the maximum building height.[26] While wind energy systems are allowed to reach a maximum of 75 feet from the ground.[27] Setback requirements for solar systems may extend three feet into the area, and systems which exceed the three feet have the option to be permitted for a conditional use.[28] The setback requirements for wind systems are separated by freestanding towers, which are required to be set back the distance of the system's height. Mounted systems must be setback according to the applicable code found in Section 15.12.000.[29] These relaxed standards for height and setback help meet the code's purpose in decreasing dependence upon non-renewable energy systems.[30]

To view the provision see Laramie, WY, Code of Ordinances § 15.14.030 (A) (1) (2017).

ENDNOTES

1   MINNEAPOLIS, MIN., ZONING CODE §§ 535.710, 535.840; SEATTLE, WASH., MUN. CODE § 23.44.046.
2   BEREN ARGETSINGER & BENJAMIN INSKEEP, STANDARDS AND REQUIREMENTS FOR SOLAR EQUIPMENT, INSTALLATION, AND LICENSING AND CERTIFICATION (Clean Energy States Alliance 2017) https://perma. cc/JT4C-KAFF. *See also* BALTIMORE, MD., CODE OF ORDINANCES § 14-0413 (2014) (adoption of the International Green Construction Code as part of the City's building code).
3   *Id.*
4   *Id.*
5   *Id.* (citing *City of Tuscon Planning and Development Services Department. "Residential Plan Review: Solar Ready Ordinance,* Ordinance No. 10549).
6   Oregon Department of Consumer and Business Services, Oregon Solar Installation Specialty Code, (Oct. 2010), https://perma.cc/74CK-ERNT.
7   ANDREW E. DESSLER, INTRODUCTION TO MODERN CLIMATE CHANGE 172 (Cambridge University Press 2012).
8   REX A. EWING & DOUG PRATT, GOT SUN? GO SOLAR: HARNESS NATURE'S FREE ENERGY TO HEAT AND POWER YOUR GRID-TIED HOME 12-13 (2d ed. 2009); DESSLER, *supra* note 7, at 172.
9   EWING & PRATT, *supra* note 8; DESSLER, *supra* note 7, at 172.
10 *See e.g.*, Wind Energy Benefits, https://perma.cc/7WH6-U44V; *See also* State Renewable Energy Resources, https://perma.cc/C564-JMJ7.
11 ANNE E. GIMMER & KAY D. WEEKS, STANDARDS FOR REHABILITATION & ILLUSTRATED GUIDELINES FOR REHABILITATING HISTORIC BUILDINGS 16 (DOI 2011), https://www.nps.gov/tps/standards/rehabilitation/sustainability-guidelines.pdf; Office of Energy Efficiency and Renewable Energy, *How Do Wind Turbines Work?*, https://perma.cc/5FRZ-3T78 (last visited Jan. 8, 2019).
12 *See* Jennifer Kuntz, *Article: A Guide to Solar Panel Installation at Grand Central Terminal: Creating a Policy of Sustainable Rehabilitation in Local and National Historic Preservation,* 10 VT. J. ENVTL. L. 315 (2009).
13 MINNEAPOLIS, MIN., ZONING CODE §§ 535.710, 535.840.
14 *Id.*
15 *Id.*
16 *Id.*
17 *Id.*
18 *Id.* § 535.750(4).
19 *Id.* § 535.840(b)(1).
20 *Id.* § 535.840 (b)(2).
21 *Id.* § 535.840(c).
22 *Id.* § 535.840(c)(2).
23 *Id* § 535.860.
24 *Id.*
25 LARAMIE, WYO., CODE OF ORDINANCES § 15.14.030.
26 *Id.* § 15.14.030 (B) (2) (a).
27 *Id.* § 15.14.030 (B) (2) (a).
28 *Id.* § 15.14.030(A)(1)(d).
29 *Id.* § 15.14.030(B)(4).
30 *Id.* § 15.14.030(A)(1)(a).

LIVE-WORK UNITS

Tyler Adams (author)
Jonathan Rosenbloom & Christopher Duerksen (editors)

INTRODUCTION

Live-Work Units (LWUs) are properties that combine residential and non-residential uses in either commercial or residentially zoned areas. LWUs are usually restricted in that they require the owner of the business to also reside in the property or vice versa.[1] LWUs were once popular pre-industrialism, but with advances in transportation, technology, and strict zoning codes that separated uses (also known as "Euclidean zoning") they almost petered out of existence by the 1950s.[2] They have since gained increasing popularity with the recent turn toward reducing carbon emissions and a desire for greater work flexibility.[3]

Local governments may separate LWUs into three use types relating to the dominance of the non-residential activity: live/work, work/live, and home occupation.[4] Many jurisdictions use the term home occupation to describe a property that is primarily used as a residence, with work being an accessory function.[5] The ordinances regulating these units typically restrict the work portion to small-scale activities and limit the numbers of employees or client visits.[6] Live/work units are similar to home occupations in that their primary use is that of a residence, but the regulations are not as restrictive. Working is permitted but is secondary to the residential component and the need to preserve the neighbor's expectations of quiet enjoyment.[7] In work/live units, the non-residential activity takes precedent over the residential activity.[8] Some local governments do not distinguish between live/work and work/live units and others incorporate two or three of the different types.

EFFECTS

Zoning ordinances and strict regulations often serve as barriers to LWUs and the numerous benefits they provide to citizens and local governments. Because of their mixed residential and non-residential nature, local governments often characterize LWUs as commercial buildings for purposes of safety regulations or prohibit LWUs in residential zones.[9] This results in an inefficient, expensive, and awkward process that is usually excessive and expensive compared to any low-risk hazard that the LWU work space may possess.[10]

If allowed to flourish, LWUs, particularly live/work and work/live units, would be able to confer numerous benefits upon a community. By eliminating the need to commute to work, car usage by an LWU owner is significantly decreased.[11] This reduces greenhouse gase emissions and vehicle miles traveled. In addition, allowing patrons to walk to their destinations reduces traffic congestion as well as the demand for parking.[12] By reducing the development of separate land parcels for different uses, LWUs also help minimize urban sprawl, again reducing a community's dependence on cars and the inefficient use of natural resources.[13] Finally, occupants of LWUs are more likely to invest in their communities due to the increased commitment to the success of the area.[14] The community and occupant are able to benefit financially from increased business activity and consequentially are in a better position to reinvest.

EXAMPLES

Grand Rapids, MI

Grand Rapids permits home occupations with a focus on preserving the character of the neighborhood and the residential quality of the home.[15] Home occupations are required to get a business license, of which there are three types: class A, home occupations that will have no impact on the surrounding neighborhood, and class B and C, those that have the potential to adversely impact the neighborhood.[16] Each class has specific characteristics, but generally they must all adhere to certain criteria. In particular, accessory structures, whether attached or detached, are not permitted to be used in connection with the home occupation. In addition, use of the home occupation cannot require exterior alterations to the dwelling, including the creation of a separate entrance.[17] Grand Rapids also permits live/work units subject to certain limitations. The unit must be on either a regional street, defined as streets that carry traffic between Grand Rapids and other communities in the region, or a major street, defined as streets that carry traffic through the city and region.[18] The ordinance specifies that residential use is to be primary to the non-residential use and that a maximum of one-half of the total area of the unit may be designated to non-residential use.[19] At least one full time employee of the non-residential activity must also reside in the unit.[20]

To view the provisions, see Grand Rapids, MI- Code of Ordinances §§ 5.9.14, 5.9.16 (Current through 2018).

Oakland, CA

Oakland, CA makes a distinction between live/work and work/live units. They define live/work units as those that accommodate both residential and non-residential activities, while work/live units are primarily nonresidential with an accessory residential area.[21] These units are permitted within the mixed-use districts and must meet certain criteria. Work/live units are divided into three types, distinguished by the maximum floor area allowed to be used for residential activities.[22] For example, type one only allows one-third of the unit to be used for residential activities while type three allows a maximum of 55%.[23] Each of the types also have distinct special requirements; type one requires that all the remaining floor space be for the non-residential activity and type two requires separate entrances for the residential and non-residential space.[24] In the live/work units, the designated floor space for residential purposes in not limited.[25] Further, the work/live and live/work units both require that the working space be regularly used by someone occupying the residential space.[26]

To view the provision see Oakland, CA-Planning Code § 17.65.040 (2016).

ENDNOTES

1   Municipal Code of Chicago § 17-9-0103.1-C (2017).
2   Marina Khoury, *Leaning Toward Live-Work Units*, Lean Urbanism Making Small Possible (May 30, 2014), https://perma.cc/XEK8-Y52U.
3   *Id.*
4   Thomas Dolan, *Live-Work Planning and Building Code Issues* 12 (Mar. 17, 2014), https://perma.cc/QR62-DBMU.
5   *Id.* at 13.
6   *Id.*
7   *Id.*
8   *Id.* at 14.
9   Khoury, *supra* note 2.
10 *Id.*
11 Delaware Regional Valley Planning Comm'n, Assessment of the Potential Role of Live/Work Development in Centers 33 (July 2014), https://perma.cc/SY9P-LGPE.
12 *Id.* at 35-36.
13 *Id.* at 36.
14 *Id.*
15 Grand Rapids, Michigan, Code of Ordinances § 5.9.14 (2018).
16 *Id.*
17 *Id.*
18 *Id.* § 5.9.16
19 *Id.*
20 *Id.*
21 *Id.*
22 *Id.* § 17.65.150 (2018).
23 *Id.*
24 *Id.*
25 *Id.* § 17.65.160 (2018).
26 *Id.* § 17.65.040 (2018).

LOCAL RECYCLING CENTERS

Kyler Massner (author)
Jonathan Rosenbloom & Christopher Duerksen (editors)

PERMIT

INTRODUCTION

Recycling has long been one of the most popular activities pursued by local communities.[1] The goal of recycling is to find ways to reuse items and materials that would otherwise be discarded into landfills. Some of the most common materials recycled include paper, plastic, and aluminum.[2] Establishing recycling centers within a municipal area helps to promote the practice of recycling as they allow for close and convenient collection points for recyclable materials.[3] This proposal would permit recycling centers in a variety of zones to help make recycling easier, convenient and efficient. Allowing recycling centers to be in or near commercial and/or residential areas can increase recycling rates by bringing the opportunities to recycle closer to people.

EFFECTS

Recycling can be both a public service and a for-profit business. Recycling removes trash and unwanted items, resulting in a benefit to the public.[4] In addition, it can provide the recycler with the opportunity to repurpose or resell the recycled materials.[5] These immediate benefits are also accompanied by reductions in greenhouse gases (GHG). Materials that could be recycled are often discarded and end up in landfills.[6] As items decay in landfills methane gas is released.[7] Methane is a powerful GHG that can be 28-36 times more efficient at trapping heat than carbon dioxide over 100 years,[8] making it a powerful and potent GHG. Reducing the amount of trash put into landfills also reduces the number and size of landfills, which in turn reduces the amount of methane that landfills produce.[9] Furthermore, recycled materials can also be more efficient than using new materials, thus reducing additional GHGs that would otherwise accrue in the manufacturing process of new materials.[10]

Fresno, CA

Fresno, CA permits the development of large recycling centers as the primary use in certain districts. Fresno requires the centers to be no smaller than three acres in size and cannot be adjacent to a residential district.[11] Allowing the primary use of a site to be a recycling center encourages development of larger recycling facilities with the ability to process larger volumes of recyclable material. In addition, Fresno allows smaller scale recycling centers to operate within the interior footprint of a business on a conditional basis.[12] This increases the availability and number of locations for recycling centers within city limits, thereby supporting wider participation due to increased proximity and greater convenience for the consumer.

Fresno also supports the use of reverse vending machines, automated machines allowing for the collection of common recyclables, such as bottles and cans, as an accessory use and allows placement nearest to a business's entrance as possible.[13] The benefit of these machines is they can operate within the interior footprint of another business. While these machines are limited in what it may accept, they require less space and labor to maintain operation. In addition, they increase the amount of recycling opportunities for consumers in various and widespread locations.

To view the provision see Fresno, CA, Code of Ordinances § 15-2750 (2018).

Big Bear Lake, CA

Big Bear Lake, CA allows both small and large recycling collection facilities.[14] Small collection centers are flexible solutions that can be responsive to the community's needs. Small facilities include reverse vending machines, kiosks, donation containers, and mobile units.[15] These small collection facilities and reverse vending machines are permitted in any General Commercial zone. Whereas enclosures and receptacles are permitted in both Commercial and Public zones.[16] Smaller units can be placed as either a permanent structure or as a mobilized unit that allows collection stations to be deployed at site specific and need specific locations, increasing user convenience, access and participation.

To view the provision see Big Bear Lake, CA, Development Code 17.02.030 (2011); Tbl. 17.35.030.A (2011); Tbl. 17.35.040.A (2011).

Madison, WI

Madison, WI created a specific use category for recycling centers permitting them in five districts.[17] Recycling centers are permitted fully in General Industrial districts and Industrial Limited districts and conditionally permitted in Traditional Employment districts, Suburban Employment districts, Suburban Employment Center districts and Employment Campus districts.[18] The ordinance establishes a sliding scale allowing recycling centers in general industrial uses where the activities of collection do not disrupt other commercial or residential functions. At the same time, it allows some recycling centers—upon conditional use—in employment and suburban districts. The conditional use permit allows increased citizen access to recycling centers, while reserving city oversight to ensure that those operations do not disrupt other community functions.

To view the provision see Madison, WI, Zoning Code, Tbl. 28F-1 (2017).

Detroit, MI

Detroit mandates that its Environmental Department prepare and implement a long-term "Green Initiatives and Sustainable Technology Plan" (GIST Plan) to establish, use and support green technology, businesses and initiatives together with both public and private partners.[19] Detroit Renewable Power (DRP) came out of the GIST Plan. DPR is an Energy-from-waste (EFW) recycling facility, that uses combustion technology to recycle municipal solid waste and turn it into renewable energy and other recovered materials.[20] Such EFWs can act as a complement to other recycling programs, thereby reducing overall environmental impacts while providing energy from a renewable resource. DRP recycles 39,000 tons of ferrous metals and nonferrous metals annually and converts 3,300 tons of municipal solid waste per day into renewable fuel.[21] DRP and General Motors (GM) were able to work together to make the GM Renaissance Center landfill-free, meaning that the Center now recycles, reuses, or converts 100% of its daily waste.[22]

See the provision at Detroit, MI, Code of Ordinances § 6-509 (2018).

ENDNOTES

1   Frank Ackerman, Why Do We Recycle: Markets Values and Public Policy 9 (1997).
2   Jon Clift & Amanda Cuthbert, Climate Change Simple Things You Can Do to Make a Difference 49-56 (2012); Shannon Tyman, *Green Cities: A-to-Z Guide, in* Recycling in Cities 367, 367-68 (Nevin Cohen ed. 2011).
3   Adam S. Weinberg et al., Urban Recycling and the Search for Sustainable Community Development Proposals 21-22 (2000).
4   *Id.* at 15-16.
5   Tyman, *supra* note 2, at 368-69; Weinberg et al., *supra* note 3, at 15-16.
6   Tyman, *supra* note 2, at 368-69.
7   Clift & Cuthbert, *supra* note 2, at 49.
8   U.S. EPA, *Understanding Global Warming Potentials*, http://perma.cc/R8A7-LBWJ (last visited Aug. 10, 2017).
9   Andrew E. Dressler, Introduction to Modern Climate Change 77 (Cambridge University Press 2012).
10 *Id.*
11 Fresno, CA, Code of Ordinances § 15-2750(C)(1)-(2) (2017).
12 *Id.* § 15-2750(B)(2) (2017).
13 *Id.* § 15-2750(A)(1) (2017); *id.* § 15-2750(A)(2) (2017).
14 Big Bear Lake, CA, Development Code ch. 17.02 (2011).
15 *Id.*
16 *Id.* at tbl. 17.35.050.A (2011).
17 Madison, WI, Zoning Code tbl. 28F-1 (2017).
18 *Id.*
19 Detroit, MI, Code of Ordinances § 6-509 (2018).
20 Damian Doerfer, *Recycling: Energy-from-waste projects are fully compatible with recycling*, Detroit Renewable Energy (2018), https://perma.cc/XAK4-74MZ (last visited May 29, 2018).
21 Doerfer, *supra* note 20; *Steam Energy from Renewable Sources*, Detroit Renewable Power (2018), https://perma.cc/3HWH-KS8X (last visited May 29, 2018); *Recycling*, Detroit Renewable Power (2018), https://perma.cc/3YV7-QCYW (last visited May 29, 2018).
22 Doerfer, *supra* note 20.

MIXED-USE

Tyler Adams (author)
Jonathan Rosenbloom & Christopher Duerksen (editors)

INTRODUCTION

Mixed-use zoning permits a complementary mix of residential, commercial, and/or industrial uses in a single district. Mixed-use zoning can take a variety of forms, but often is categorized as one of three types: vertical mixed-use, horizontal mixed-use, and mixed-use walkable.[1] Vertical mixed-use allows for a combination of different uses in the same building and most frequently the non-residential uses occupy the bottom portion of the building, with the residential on top.[2] Horizontal mixed-use allows distinct uses on separate parcels to be combined in a particular area or district. This helps avoid the complexities of combining uses that may have different safety or regulatory requirements in a single building.[3] Mixed-use walkable combines vertical mixed-use and horizontal mixed-use, thus creating an area containing mixed-use buildings as well as distinct single-use buildings in close proximity to each other.[4]

Prior to the rise of the automobile and modern zoning practices, mixed-use developments were the norm.[5] Since the rise of classic Euclidean Zoning, use segregation has been the norm and integrated land uses have been relatively rare.[6] The emergence of sustainability and walkability as important factors in community development has led to a resurgence of mixed-used zoning.[7] Implementation of mixed-use zoning has evolved to include more than just permitting mixed-use developments in certain districts. Local governments are now creating mixed-use districts. This allows for a more widespread integration of uses and the development of increasingly cohesive and efficient communities.

EFFECTS

Mixed-use zoning can provide several important benefits, including:

- Reducing combined housing and transportation costs for households by providing diverse housing options and alternatives to automobile travel;

- Creating cohesive, yet diverse, neighborhoods with increased economic and cultural opportunities, contributing to greater livability and a healthier local economy;

- Encouraging healthier lifestyles by creating a pattern of development in which biking and walking are part of everyday travel behaviors;

- Reducing vehicle miles traveled, dependence on fossil fuels, and associated greenhouse gas emissions;

- Reducing the costs of delivering public services by encouraging infill and redevelopment in areas with existing infrastructure;

- Providing a more compact development pattern that helps preserve open space and natural resources elsewhere in the community or region;

- Encouraging a more sustainable transportation system over the long term by creating viable options for people to get to destinations by multiple modes of transportation;

- Reducing reliance on building new roadways or widening existing roadways to meet transportation needs as a community and region continues to grow; and

- Taking advantage of and facilitating public investments in transit infrastructure, enabling more efficient servicing of community and regional transportation needs.[8]

When implementing mixed-use zoning, municipalities should consider how to mitigate potential adverse impacts related to mixed-use developments and buildings. Such negative impacts may include increased traffic, differing parking needs for residential and commercial uses, and insufficient existing infrastructure.[9]

EXAMPLES

Baltimore, MD

In 2017, the city of Baltimore's most recent zoning code update went into effect. The updated code included the addition of new mixed-use zoning districts aimed at boosting the economic development of the City as well as preserving the existing character.[10] For example, an industrial mixed-use zoning district was added, intending to encourage the reuse of older buildings for light industrial use and other non-industrial uses.[11] An industrial use must account for at least 50% of the total ground floor area of all buildings on the lot or a use other than residential must account for at least 60% of the total ground area. Also, the addition of Rowhouse Mixed-Use Overlay districts as well as Detached Dwelling Mixed-Use Overlay districts were intended to address those areas where a mixed-use environment was desired in rowhouse or detached dwelling developments.[12] The overlay districts are directly tied to the underlying rowhouse or detached dwelling district in order to preserve the existing character and development of the neighborhoods.[13] Commercial and non-residential uses are restricted to those that are compatible with the existing residential use.[14] For instance, detached dwelling mixed-use districts limit the permitted non-residential to the ground floor of the dwelling and only four types of uses are permitted.[15]

To view the provision see Baltimore, MD, City Code, Art. 32 §6-201 (2017).

St. Anthony, ID

The city of St. Anthony established mixed-use zones to provide for commercial aspects in neighborhood centers.[16] These centers should include a mix of land uses that are "located together either vertically or horizontally within the same building as well as a mix of individual residential and commercial buildings in close proximity."[17] There are two types of mixed-use districts. The low intensity mixed-use district (MU1) has an allowable housing unit density of up to 8 units per acre and is intended to have a more residential style rather than commercial.[18] The moderate intensity mixed-use district (MU2) has a permitted housing unit density of 16 units per acre with a special use permit and is more commercial in style.[19] In both districts, a principle building is required to have its main entrance accessible through a public sidewalk or a private sidewalk publicly accessible through a public use easement in order to encourage pedestrian-oriented developments.[20]

To view the provisions see St. Anthony, ID, Municipal Code §§ 17.06.090-17.06.120 (2015).

ENDNOTES

1   Howard Blackson, *Don't Get Mixed Up on Mixed-Use*, PlaceMakers (Apr. 4, 2013),https://perma.cc/SPW9-Q3DH.
2   *Id.*
3   *Id.*
4   *Id.*
5   *Mixed-Use Development 101: The Design of Mixed-Use Buildings*, Urban Land Institute (Aug. 30, 2011), https://perma.cc/3ZWU-5UPN.
6   *Id.*
7   *Id.*
8   Jill Grant, *Mixed Use in Theory and Practice*, 68 J. Am. Planning Ass'n 71, 72-73 n.1 (2007); Adrienne Schmitz & Jason Scully, Creating: Walkable Places: Compact Mixed-use Solutions 21-23 (2006).
9   Grant, *supra* note 8, at 70-80 n.1.
10   Archana Piyati, *Baltimore's New Zoning Hoped to Boost More Mixed-Use Development*, Urban Land (Aug. 8, 2017), http://perma.cc/YPL9-TM3D.
11   Baltimore City Code, Art. 32 § 11-203 (2017).
12   *Id.* §§ 12-208, 12-209.
13   *Id.*
14   *Id.*
15   *Id.* § 12-1105.
16   St. Anthony Municipal Code § 17.06.090 (2015).
17   *Id.*
18   *Id.* § 17.06.110.
19   *Id.* § 17.06.120.
20   *Id.* §§ 17.06.110-17.06.120.

Renewable Energy for Historic Buildings

Alec LeSher (author)
Jonathan Rosenbloom & Christopher Duerksen (editors)

INTRODUCTION

Technological advancements in wind and solar energy have increased the ability to incorporate alternative energy systems in historically significant buildings. Local governments can draft ordinances that require compatible installation of solar panels and wind turbines in designated historical districts. An ordinance that provides the structure for incorporating renewable energy systems into historic districts should primarily focus on maintaining the aesthetic of the historic building.[1] A proper ordinance should include, but is not limited to, setback requirements, placement requirements, design standards, and whether a certificate of appropriateness is required.[2]

Ordinances allowing solar energy systems should permit solar panels that are directly exposed to the sun, ideally on the roof, and that are low profile, so as not to compromise the historic character of the property.[3] This may mean requiring the use of flush mounted solar panels that are not visible from the street. Alternatively, solar panels can be placed on the roof that does not face the public right of way. An ordinance can also require solar panels to be placed at lower angles or flush to the roof to keep them out of the public's sight. If the roof of a historic building is a defining feature of the property, solar panels may be placed in secondary locations, such as pole mounted arrays behind the building or on adjacent lands, so long as they remain hidden from the public or do not interfere with the historic value of the building.[4]

Ordinances allowing wind turbines, should also be drafted to ensure that installation does not diminish the historic character of the property.[5] In order to be most effective, wind turbines must typically be taller than the structures around them.[6] Due to their highly visible nature, wind turbines may have difficulty finding an efficient location in historic districts, where they are not diminishing the historic value of the properties.[7]

One common concern with adding renewable energy systems is that the systems are incompatible with the historic nature of the buildings and sites.

The Secretary of the Interior's Standards for the Treatment of Historic Properties (the Standards) offers four approaches for the preservation, rehabilitation, restoration, and reconstruction of historically significant sites and buildings.[8] The Standards are supported by Interior's Guidelines, which "offer general design and technical recommendations to assist in applying the Standards to a specific property."[9] One prominent requirement found in the Standards is that alterations to historic properties should not diminish the character of a historic site or building.[10] It is commonly believed that this provision and new environmentally-friendly technologies are incompatible because the inclusion of renewable energy systems may have a substantial effect on the character of the structure. Given the technological advances in renewable energies, this belief is, in many cases, incorrect. Municipalities have successfully incorporated renewable energy systems into historic buildings in ways that comply with the Standards and maintain the historic character of the building.[11]

EFFECTS

Ordinances for the compatible installation of solar panels and wind turbines in designated historic districts will ensure that renewable energy systems are installed in a manner that is consistent with the character of historic properties.[12] There are many benefits of solar and wind energy that have attributed to their rising popularity. First and foremost, solar and wind energy are inexhaustible renewable resources, resources that will not be depleted upon use.[13] This may lead to less dependence on non-renewable resources, such as coal and natural gas.[14] Non-renewable fossil fuel sources of energy also produce far more greenhouse gas (GHG) emissions, which is a primary cause of climate change.[15] According to the Environmental Protection Agency: "Increasing the use of renewable energy is one of the most effective ways to quickly reduce [GHG] emissions."[16]

Another advantage of renewable energy systems is that they reduce air pollutants that adversely affect human health and the environment.[17] Emissions from burning fossil fuels adversely affect human health by contributing to respiratory illnesses, smog, and global climate change.[18] In sum, integrating energy improvements into historic buildings will reduce energy consumption, mitigate the effects of climate change, and promote human health, all while preserving the aesthetic and cultural value of the historic area.[19]

Bay City, MI

The Bay City Michigan Code approves of solar systems in historic districts and provides guidelines for the installation of solar panels on rooflines.[20] The historic district commission must approve any solar system in a historic district.[21] To be approved, solar panels must not damage or obscure character-defining features on a historic building and should not alter or obscure a historic roofline.[22] When installing a solar system on a pitched roof, the panels may only be attached on the roof-side that is visible from public streets if there is no other location that is not visible from public streets.[23] When installing solar panels on flat roofs, panels should not be attached to parapet walls that are clearly visible from public streets and should be set back so that they are not clearly visible from the public street.[24] If mounting solar panels on a roof is not feasible, the ordinance also allows for pole mounted solar arrays to be constructed in locations not clearly visible from the public street.[25] Lastly, the historic district commission may require the solar systems be painted in way that blends in with the surrounding area.[26]

ADDITIONAL EXAMPLES

Houston, TX, Code of Ordinances §§ 33-241.1 (c) (4), 33-237 (2015) (permitting construction of solar panels in historic districts on the front part of the roof upon the grant of a certificate of appropriateness, but if on the rear part of the roof no certificate is needed).

East Greenwich Twp., NJ, Code of Ordinances § 16.72.030 (2010) (allowing solar panels and wind turbines to be constructed in a historical district if the use conforms to setback and height restrictions).

Breckenridge, CO, County Code § 9-1-19-5A:E (2013) (allowing for the installation of solar panels only where it will not be detrimental to the character of the historic building; requiring solar panels to be installed in a way so as not to be visible from a public street).

Newberry, FL, Code of Ordinances § 11.11.2.1 (2015) (providing that land development regulation administrator may issue a certificate of appropriateness for the installation of solar panels; solar panels must not damage character-defining features, must be located on an addition to the historical building, and should blend into the surrounding features).

National City, CA, Code of Ordinances § 18.30.300 (D) (2012) (providing that solar panels should be installed in locations that are not visible from a public way, run parallel to the original roofline, set back to minimize visibility, and should be of similar color to the roof).

To view the provision see Bay City, MI, Code of Ordinances § 122-625 (2012).

Las Cruces, NM

The Las Cruces Code states: "Enhancing the energy efficiency of a historic building is important."[27] The goal of this section is to allow the installation of solar panels in a way that does not damage the integrity of historic sites.[28] The Code complies with the Standards by insuring that solar panels do not alter the historic character of the building. The solar panels must preserve the aesthetic of the building and be reversible, that is, capable of sending excess energy back to the grid.[29] When installed, they should not be highly visible from the public right of way if possible.[30] This means that panels must be placed as close to the roofline as possible, if not completely flush. On a flat roof, solar panels must be setback from the edge, and angled so that they cannot be seen from the street.[31] All panels and mounting equipment must be colorized to blend into the structure.[32] If a solar array is not placed on the roof, it must be placed in an area with limited or no-visibility from the public street.[33] This can be done by planting vegetation around the panels or by using of a manufactured screen.[34]

To view the provision see Las Cruces, NM, Land Development Code § 38-49.2 (M) (2015).

ADDITIONAL RESOURCES

Office of Energy Efficiency and Renewable Energy, *Renewable Electricity Generation*, https://perma.cc/KYX5-D24E (last visited June 6, 2018).

Grow Solar Local Government Solar Toolkit: Planning, Zoning, and Permitting, https://perma.cc/LDQ6-5UV6 (last visited June 6, 2018).

National Renewable Energy Laboratory, *Implementing Solar PV Projects on Historic Buildings and in Historic Districts*, Sept. 2011, https://perma.cc/46JH-2E4V.

ENDNOTES

1   *See* Anne E. Gimmer & Kay D. Weeks, Standards for Rehabilitation & Illustrated Guidelines for Rehabilitating Historic Buildings vii (DOI 2011) , https://www.nps.gov/tps/standards/rehabilitation/sustainability-guidelines.pdf.

2   *See* Energy Efficiency, Renewable Energy, and Historic Preservation: A Guide for Historic District Commissions 34-35 (no date), https://perma.cc/R4MD-VADB (last visited June 6, 2018); *see also* Great Plains Institute, *Grow Solar Local Government Solar Toolkit: Planning, Zoning, and Permitting*, https://perma.cc/LDQ6-5UV6 (last visited June 6, 2018).

3   National Park Service, *ITS 52: Interpreting The Secretary of the Interior's Standards for Rehabilitation* (Aug. 2009), https://perma.cc/AJY2-UDPW; Office of Energy Efficiency and Renewable Energy, *Solar Technology Basics* (Aug. 16, 2013), https://perma.cc/2TEB-L33U.

4   National Park Service, *supra* note 3.

5   Gimmer & Weeks, supra note 1, at 16; Office of Energy Efficiency and Renewable Energy, *How Do Wind Turbines Work?*, https://perma.cc/A8X6-64F9 (last visited June 6, 2018).

6   Energy Efficiency, *supra* note 2, at 35.

7   *Id.*

8   The four Standards approaches "are regulatory for all grant-in-aid projects assisted through the national Historic Preservation Fund. The Standards for Rehabilitation, codified in 36 CFR 67, are regulatory for the review of rehabilitation work in the Historic Preservation Tax Incentives program." Otherwise, the Standards are advisory only. *The Secretary of the Interior's Standards*, National Park Service, https://perma.cc/Y4SS-8SPB (last visited June 6, 2018).

9   The Guidelines are advisory only. *Id.*

10 *Id.*

11 *See e.g.*, Bay City, Mich., Code of Ordinances §§ 64-11 (2005), 122-625 (2012); Las Cruces, N. Mex., Code of Ordinances §§ 2-547 (2017), 38-49.2 (2015).

12 *See Renewable Energy Ordinance Framework*, Delaware Valley Regional Planning Commission, https://perma.cc/829U-CLNM (last visited June 6, 2018); *see also* Monroe County, Wis., Code of Ordinances § 47-675 (2006).

13 *See e.g.*, U.S. Department of Energy, National Renewable Energy Laboratory, *Wind Energy Benefits*, (Jan. 2015), https://perma.cc/G8WH-QB2Q; *see also* U.S. EPA, *State Renewable Energy Resources*, Energy Resources for State and Local Governments, https://perma.cc/6NXG-A977 (last visited Jan. 2, 2019).

14 Amy L. Stein, *Renewable Energy Through Agency Action*, 84 U. Colo. L. Rev. 651, 653 (2013).

15 U.S. EPA, *Human Related Sources and Sinks of Carbon Dioxide*, https://perma.cc/6SAL-DZF9 (last visited May 31, 2018) (stating that the process of creating energy is the single largest source of greenhouse gas emissions in the United States).

16 *Stein, supra* note 14, at 663-64.

17 *See, e.g., supra* note 13.

18 *Stein, supra* note 14, at 663-64; *see also* Michael B. Gerrard & J. Cullen Howe, *Global Climate Change: Legal Summary*, ST038 ALI-ABA 831 (2012).

19 *See* Jennifer Kuntz, *Article: A Guide to Solar Panel Installation at Grand Central Terminal: Creating a Policy of Sustainable Rehabilitation in Local and National Historic Preservation*, 10 Vt. J. Envtl. L. 315 (2009).

20 Bay City, MI, Code of Ordinances § 122-625 (2012).

21 *Id.* § 122-625(1).

22 *Id.* § 122-625 (2)(a).

23 *Id.* § 122-625 (2)(c)(1)(a).

24 *Id.* § 122-625 (2)(c)(2)(a).

25 *Id.* § 122-625 (2)(b).

26 *Id.* § 122-625 (2)(e).

27 Las Cruces, N.M., Land Development Code § 38-49.2 (M) (2015).

28 *Id.*

29 *Id.*

30 *Id.*

31 *Id.*

32 *Id.*

33 *Id.*

34 *Id.*

SOLAR ENERGY SYSTEMS AND WIND TURBINES BY-RIGHT

Kerrigan Owens (author)
Jonathan Rosenbloom & Christopher Duerksen (editors)

INTRODUCTION

These ordinances seek to increase renewable energy, specifically wind and solar, by permitting solar energy systems and wind turbines by-right in certain zoning districts.[1] Currently, some local government codes contain districts that not only fail to protect solar energy systems and/or wind turbines, but also explicitly or implicitly (through height, setback, and other requirements) prohibit them in some neighborhoods (see Height and Setbacks for Wind and Solar brief for an ordinance removing and altering these restrictions).[2] Local governments may enact ordinances to permit solar energy systems and/or wind turbines by-right in some zoning districts. Doing so eliminates zoning barriers and increases efficiency of installation.[3] When considering these ordinances, local governments should address use restrictions, which districts should allow solar energy systems and wind turbines by-right, height and set back requirements, design, quantity, minimum and maximum energy output of installations, landscaping requirements, permitting, and other concerns specific to the district. For purposes of permitting, local governments may consider permitting installations by-right or upon conditional permit approvals or some other mechanism for review, making it clear that the installations are permitted upon approval.

At the state level, California's legislature has stated that any type of covenant or zoning restriction that prohibits solar energy systems are void and unenforceable.[4] Although the statute still allows for zoning restrictions, the CA Civil Code outlines that the restrictions cannot result in the addition of $1,000 or more to the cost of solar installation nor can the restrictions limit the efficiency of the system by 10% or more.[5] Florida's legislature passed similar legislation in 2008, expressly prohibiting ordinances which prohibit solar collectors.[6] The statute further states that there can be no prohibition on solar collectors in the form of covenants, deeds, or other similar agreements.[7] The statute allows the HOA or other home owner entity to determine where the collectors may be placed.[8]

EFFECTS

Based on a 2017 Energy Information Agency (EIA) study, 29 percent of U.S. greenhouse gas emissions come from electricity, mostly from the burning of coal and natural gas.[9] Comparatively, renewable energy sources produce little to no emissions.[10] Moreover, in 2016 there were over 500 factories associated with renewable energy sources in the U.S. producing more than $13.0 billion in revenue.[11] The benefits of renewable energy sources could be greatly enhanced by permitting solar panels and/or wind turbines by-right within certain districts. A study produced by the National Renewable Energy Laboratory estimated that the United States could provide up to 80 percent of electricity from renewable sources by 2050.[12] The benefits of doing so include fewer emissions, improved public health, and less strain on water resources.[13] One study from Harvard University estimated that the public health costs of coal are $74.6 billion a year.[14] In contrast, solar and wind energy have fewer negative health impacts in terms of air or water pollution.[15] Shifting toward renewable resources also diversifies the energy pool, lessening the need for imported energy sources. By permitting wind and/or solar by-right within certain districts, local governments would facilitate the reduction of greenhouse gases and a move toward energy independence. By permitting wind and/or solar by-right within certain districts, local governments would facilitate the reduction of greenhouse gases and a move toward energy independence through a system that has the potential to pay for itself. Because of their decentralized nature, such systems also enable the resilience of energy infrastructure.

EXAMPLES

Bedford, NY

The Town of Bedford, New York allows solar energy collectors by-right in all districts.[16] The solar energy collectors are designated as accessory buildings and structures and must comply with the building requirements and setbacks for accessory buildings and structures.[17] However, to maximize the efficiency of solar energy collectors, the collectors do not have to follow standard height limitations as set in the code.[18] The ordinance states that the solar energy collector does not have to comply with the maximum height standard, but the collector must not be fifteen feet above the roof nor can the collector cover more than 10% of the total roof area.[19]

To view the provision see Bedford, NY, Zoning Code, §125-20 and §125-27 (2017).

Stoughton, WI

The City of Stoughton, Wisconsin in early 2018 enacted a local ordinance that allows solar energy production, by-right, in every zoning district.[20] Permitted districts include Agricultural, Residential, Office, Business, Industrial, and Institutional.[21] The solar systems must comply with regulations and limitations set forth for installations in the respective districts, including height and setback, but are allowed by-right within these districts.[22] The City also mandates that wind systems will be allowed in each district listed above, but on a conditional basis conforming to setback and height requirements.[23] The setback requirement for the wind systems require them to be back from the property line at least 1.1 times their height.[24] The height requirement falls into three categories; for properties under two acres, the system may be up to 60 feet, for properties between two and five acres the system may be up to 100 feet, and for properties over five acres the system may be up to 150 feet.[25]

To view the provision see Stoughton, WI, Code of Ordinances, § 75-105 (2018).

Hartford, CT

Hartford permits building-mounted solar and roof-mounted wind energy systems as an accessory structure by right within all jurisdictions, subject to use-specific regulations.[26] Freestanding wind and solar energy systems, as well as solar-collecting canopies over parking lots, are permitted as accessory structures by right in a limited number of jurisdictions and are likewise

---

ADDITIONAL EXAMPLES

Bethany Beach, DE, Zoning Code, § 484 (2010) (permitting the use of solar energy systems in all districts subject to minimal regulations).

Cleveland, OH, Zoning Code, § 354A (2009) (permitting wind energy facilities by right).

Minneapolis, MN, Zoning Code §§ 535.690, 535.820 (2007) (permitting the production of both wind and solar energy in all zoning districts).

West Lake Hills, TX, Code of Ordinances § 22.03.009 (2017) (allowing the use of solar energy devices in all zones).

Schaumburg, IL, Code of Ordinances § 154.70 (2018) (permitting small scale wind and solar as accessory use in most zoning districts).

subject to specific use regulations.[27] Each renewable energy system in the ordinance is regulated by categories such as quantity, height, location, output or capacity, height, setback, installation method, and building material.[28] A few other categories are tailored for individual systems, such as the permitted location of free-standing windmills within the City.[29]

Building mounted solar-energy systems may include photovoltaic or hot water solar energy systems, but are not explicitly limited to these two types.[30] These solar instruments are defined as a: "system affixed to or an integral part of a principal or accessory building," and may be placed in or on roofing materials, windows, skylights, and awnings.[31] The code allows solar-collecting instruments on parking lot canopies to be both free standing structures over previously uncovered lots in addition to covering the top story of a parking structure (i.e. a parking garage).[32] One roof-mounted wind system is allowed for every 750 square feet of combined roof surface per zoning lot, and may be installed on buildings with a minimum height of 40 feet or four stories.[33] In the districts which allow free-standing wind turbines, the structures must be within 1,000 feet of the Connecticut River or an interstate, and must be set back 1.1 times its height from adjoining structures, walkways, and property lines.[34]

To view the provision see Hartford, CT, Hartford Zoning Regulations § 4.20 (2018).

ADDITIONAL RESOURCES

Local Renewable Energy Benefits and Resources, Environmental Protection Agency (Apr. 19, 2019), https://www.epa.gov/statelocalenergy/local-renewable-energy-benefits-and-resources#one.

Khagendra P.Bhandari et al., *Energy payback time (EPBT) and energy return on energy invested (EROI) of solar photovoltaic systems: A systematic review and meta-analysis*, Renewable and Sustainable Energy Reviews Vol. 47, 133-41 (July 2015), https://www.sciencedirect.com/science/article/pii/S136403211500146X?via%3Dihub.

ENDNOTES

1   Greer Ryan, Throwing Shade: 10 Sunny States Blocking Distributed Solar Development. (Center for Biological Diversity 2016), https://perma.cc/9ZGR-PPNQ.
2   *Id.*
3   *Id.*
4   Cal. Civil Code § 714 (2014).
5   *Id.*
6   Fla. Stat. § 163.04 (2008).
7   *Id.*
8   *Id.*
9   Energy Information (EIA), How Much of the U.S. Carbon Dioxide Emissions are Associated With Electricity generation? (2017).
10   *Id.*
11   Ryan Wiser & Mark Bolinger, *Wind Technologies Market Report* (U.S. Department of Energy YEAR?).
12   *Estimating Renewable Energy Economic Potential in the United States: Methodology and Initial Results,* (NREL 2016), https://perma.cc/V38T-723F.
13   *Benefits of Renewable Energy Use,* Union of Concerned Scientists (Dec. 20, 2017), https://perma.cc/26XQ-ZVSW.
14   *Id.*
15   *Id.*
16   Bedford, NY, Zoning Code § 125-27(B).
17   *Id.* at §125-27 (C), (D).
18   *Id.* at §125-20.
19   *Id.* at §125-20(A), (B).
20   Stoughton, Wis., Code of Ordinances § 78-105(1)(a)(5), (2)(a)(5), (2)(b)(5), (2)(f)(5), (2)(g)(5), (3)(a)(6), (3)(b)(5), (4)(a)(6), (4)(b)(5), (4)(c)(5), (5)(a)(5), (5)(b)(5), (5)(c)(5), (6)(a)(3)(a).
21   *Id.*
22   *Id.* at § 78-206(10)(c).
23   *Id.* at § 78-105(1)(a)(6), (2)(a)(6), (2)(b)(5), (2)(f)(6), (2)(g)(6), (3)(a)(7), (3)(b)(6), (4)(a)(7), (4)(b)(6), (4)(c)(6), (5)(a)(6), (5)(b)(6), (5)(c)(6), (6)(a)(3)(b).
24   *Id* at § 78-206(10)(b).
25   *Id* at § 78-206(10)(c).
26   Hartford, CT, Hartford Zoning Regulations § 4.20.1 (2018).
27   *Id.*
28   *Id.* § 4.20.6(A-E).
29   *Id.* § 4.20.6(D).
30   *Id.* § 4.20.6(A).
31   *Id.*
32   *Id.* § 4.20.6(C).
33   *Id.* § 4.20.6(E).
34   *Id.* § 4.20.6(D).

TINY HOMES AND COMPACT LIVING SPACES

Kerrigan Owens (author)
Jonathan Rosenbloom & Christopher Duerksen (editors)

INTRODUCTION

Since 2007, "tiny homes" have become increasingly popular in the United States.[1] Appendix Q to the 2018 International Residential Code ("IRC") defines a tiny house as a dwelling that has a floor area of 400 square feet or less, excluding lofts.[2] In addition, the IRC model definition of a tiny house requires a minimum ceiling height for habitable spaces (not less than 6 feet 8 inches), as well as overall compliance with relevant emergency escape and rescue openings specifications.[3] A local government can permit permanent occupancy of "tiny homes" by amending its zoning code to specifically permit use of land for tiny homes as a primary or secondary detached dwelling unit in existing residential districts.[4]

Generally, an ordinance permitting tiny homes as permanent dwellings requires a match between the definition of "detached second dwelling unit" and/or "single/family home" under permitted land uses in the local zoning ordinance and the specifications set out for tiny homes.[5] Alternatively, minimum specifications for homes, such as square footage, may be lowered to permit tiny homes. Local governments may permit tiny homes by-right or upon: 1) conditional use permit by the planning commission; 2) special exception by the zoning board; or 3) building permit approved and issued by the building inspector.[6] The existing definitions of permitted uses in the development code can be used to determine how and whether a tiny home should be permitted.[7] In addition to the content of the permitted use and required structure specifications, any community considering an ordinance permitting tiny homes as permanent dwellings should provide for:

- Zones Where Allowed: The city should identify specific districts where the tiny homes are permitted, depending on how they are defined in the code.[8]

- Standards Applied to Tiny Homes: All applicable zoning standards should apply to tiny homes if they are being regulated as single-family units. If treated as a separate type of land use, then any exceptions to the previous sentence should be explicitly noted and included in the code.[9]

- Minimum Dwelling Size/Occupancy: The IRC requires every dwelling unit to have "at least one habitable room that shall have not less than 120 square feet (11 [meters squared]) of gross floor area."[10]

EFFECTS

Permitting permanent occupancy of "tiny homes" in residential zones is a particularly cost-effective way to maximize affordable housing options within an existing community. To begin, the cost of constructing tiny homes is substantially cheaper than the cost of constructing traditional housing.[11] In addition, because second dwelling units can be constructed on land that is already developed and has access to existing utilities, infrastructure, and other community services, the city and/or the developer is able to avoid almost all costs other than those associated with extending existing utilities to service the additional unit.[12] Bypassing these costs can help the community increase the supply of affordable housing units by maximizing use of existing land and lower housing production costs.[13] In addition, community advisors can help further facilitate development by making sure the application and approval process for building a detached tiny home or second dwelling is not so onerous as to deter the interest of existing single-family residential districts.[14]

Amending a city's zoning ordinance to permit tiny homes is not only one of the most cost-effective options for providing affordable housing in an existing community, but also one of the swiftest options available.[15] Typically, an endeavor to create affordable housing is costly because it requires large-scale construction and development, which can be a slow process and may require a variety of government subsidies and approvals.[16] Tiny homes offer a uniquely economical approach to increasing the availability of affordable housing because the local government must merely amend the definitions within existing zoning ordinances in order to initiate development. The cost of tiny homes is almost entirely shifted to the private market.[17]

Three of the most common concerns that arise when permitting the use of tiny homes in residential zones are the impact on the character of the community, the effect on the property value of adjacent lots, and the potential burden on parking.[18] As illustrated below, there are various ways to draft ordinances to help the community control the impact that permitting tiny homes may have on an existing community.

EXAMPLES

Spur, Texas

The City Council of Spur, Texas proclaimed it was the first tiny home friendly town in the nation in July of 2014.[19] The Council realized the popularity of tiny homes due to the economic feasibility and freedom associated with them.[20] Additionally, the town saw tiny homes as a way to facilitate local growth.[21] The ordinance generally permits tiny homes in the city by right.[22] However, individuals need a variance to have a tiny home in seven of the city's subdivisions.[23] Additionally, all builders must acquire a permit prior to building and the permit application must contain descriptions of the materials being used, plat or blueprints which identify connection to utilities, and the legal property description.[24]

Tiny homes on wheels are not permitted, unless the wheels are removed and it is tied to a foundation.[25] Otherwise, the tiny home must have a cement footing of at least six inches.[26] Additionally, the tiny home must be connected to the city utilities, and compositing toilets are not permitted.[27] To qualify as a tiny home, it must have less than 900 square feet of living space.[28] However, there is not a minimum requirement of square feet living space.[29] With this tiny home ordinance, the town is able to regulate what is built, while still allowing a viable option for affordable housing.

To view the provision see Spur, TX, Ordinance 677.

Fresno, California

The Fresno, California zoning code permits use and construction of tiny homes as permanent dwellings and provides for second dwelling units, backyard cottages, and accessory living quarters as a specific type of permitted use in residentially zoned lots.[30] The purpose of Fresno's land use ordinance is to permit second dwelling units as an accessory use to single-unit dwellings and set standards for those second dwelling units that ensure compatibility of new units with the character of existing neighborhoods.[31] The Fresno ordinance requires a minimum lot size of 6,200 square feet in order to obtain a permit for a second dwelling unit, and a minimum lot size of 6,000 square feet for backyard cottages.[32] The city's district standards specifically prohibit even minor deviations and/or variances when meeting minimum lot sizes.[33] Similarly, the Fresno ordinance also implements a maximum floor area requirement for second dwelling units and limits the maximum number of accessory dwelling units that can be located in residentially zoned lots, stating that there can only be one in addition to the primary dwelling.[34]

One notable aspect of the Fresno ordinance is the six-component definition of a tiny house it added to its code.[35] Within this definition, the city heightened specific safety requirements by referencing independent product safety standards.[36] In order to preserve the character of the neighborhood and protect the integrity of the community, the Fresno ordinance requires that there be an all-weather surface path to the second dwelling unit from the street frontage.[37] Requiring a path from the accessory unit to the street front seems to add an element of permanence to the structure's character, which is often thought to preserve those romantic qualities of the neighborhood that render it "residential."[38] The Fresno ordinance goes further to permit use of a tiny home on wheels as a second dwelling unit.[39] However, in line with the purpose of this ordinance to permit permanent residence of second dwelling units, the language of the code qualifies this feature by including that in order to be in compliance, the unit, if it has wheels, must be skirted and not designed to move on its own.[40]

To view the provision see Fresno, CA, § 15-2754, Second Dwelling Units, Backyard Cottages, and Accessory Living Quarters.

ADDITIONAL RESOURCE

2018 International Residential Code for One- and Two-Family Dwellings, Appendix Q, https://perma.cc/VRL4-N666.

ENDNOTES

1 Devon Thorsby, *The Big Impact of Tiny Homes: How Little Houses Are Changing Real Estate*, U.S. News & World Report (Aug. 5, 2016), https://realestate.usnews.com/real-estate/articles/the-big-impact-of-tiny-homes-how-little-houses-are-changing-real-estate.

2 Tiny Houses, 2018 International Residential Code, App. Q, https://perma.cc/VRL4-N666 (giving guidance on requirements for tiny houses on foundations).

3 *Id.*

4 *Living Tiny Legally, Part 1*, a three-part docuseries by Tiny House Expedition, and educational Resource for Tiny House Advocates and Policy Makers, https://www.youtube.com/watch?v=ZfLAKgJGc2g

5 *Id.*

6 Rockingham Planning Commission, Accessory Dwelling Unit Model Ordinance (Sept. 29, 2016), available at http://www.rpc-nh.org/application/files/4314/7524/1047/RPC_ADU_Model_Ordinance_20160929.pdf.

7 Homelessness and Housing Toolkit for Cities, "Tiny Homes" As Permanent Housing - Zoning and Code Limitations 23 (Ass'n of Washington Cities & Municipal Research Ctr. 2017) https://perma.cc/N5SW-P467.

8 *Id.*

9 *Id.*

10 *Id.* (noting most tiny homes tend to be larger than 120 square feet).

11 David Eisenberg, *2018 Tiny Houses Appendix Q in International Residential Code*, Northwest EcoBuilding Guild, https://perma.cc/D8JJ-L496 (citing statistics from U.S. Census Bureau for proposition that a steady increase in the cost of building an average home in the United States in conjunction with a stagnant level of average annual income in the past several years has contributed to willingness to entertain "tiny houses" in the IRC).

12 *Living Tiny Legally, Part 1, supra* note 4.

13 *Id.*

14 *Id.* (suggesting streamlining approval processes for necessary permits in order to make the application process less burdensome to interested property owners).

15 *Id.*

16 *Id.*

17 *See id.*

18 Rodney L. Cobb & Scott Dvorak, American Planning Ass'n, Accessory Dwelling Units: Model State Act and Local Ordinance (2000).

19 *Welcome to Spur,* SpurFreedom, https://perma.cc/NHD8-JTNM (last visited May 23, 2018).

20 *Id.*

21 *See The Most Complete Resource on the Tiny House Movement in Spur, Texas,* Spur Tiny Houses, spur.lifeonthe.cloud (last visited Mar. 14, 2018).

22 *Spur Tiny House Packet,* www.spurfreedom.org/wp-content/uploads/2016/05/Tiny-House-Packet.pdf, 9 (last visited Apr. 15, 2018).

23 Spur, Tex., Ordinance 677, § III(1)-(2) (amended Mar. 15, 2016).

24 *Id.* § III(3)-(5).

25 *Id.* § IV(1).

26 *Id.*

27 *Id.* § IV(E).

28 *Id.* § II(1)(H).

29 *See id.*

30 Fresno, Cal, Mun. Code § 15-2754.

31   *Id.*
32   *Id.*
33   *Id.* at 15-2754(C).
34   *Id.*
35   California Tiny House, New Zoning Code, https://perma.cc/9YQU-WKLJ (providing a complete copy of the relevant sections of the City of Fresno Development code as related to accessory dwelling units).
36   *See id.*
37   Fresno, Cal, Mun. Code § 15-2754.
38   *See Living Tiny Legally, Part 1, supra* note 4 (speaking about the importance of preserving residential qualities of the neighborhood, generally indicating that the appearance of "permanency" is one of the most common ways to minimize impact of tiny homes on the existing aesthetic of the community).
39   *See* Santa Rosa, Cal., City Code § 20-42.130, *but see* Fresno, Cal, Mun. Code § 15-2754.
40   *Id.*

Part 2:
CREATE INCENTIVES

DENSITY BONUS FOR INSTALLATION OF SOLAR ENERGY SYSTEMS

Gabby Gelozin (author)

Darcie White, Sara Bronin, Brett DuBois, & Jonathan Rosenbloom (editors)

INTRODUCTION

Solar power is increasingly popular and affordable for residential use.[1] A majority of the residential installations in recent years have been on middle-class homes.[2] Providing density bonuses in exchange for installing solar energy systems encourages the growth of green energy while mitigating sprawl. Density bonuses are an "incentive-based tool that permits a developer to increase the maximum allowable development on a site in exchange for either funds or in-kind support for specified public policy goals."[3] The most popular type of density bonus is an of-right incentive.[4] This type of incentive grants the density bonus if the developer provides a set amount of the specified public interest, such as installing solar panels.[5] Local governments can apply this type of incentive to solar power by granting density bonuses to developers if a certain portion of the new development gets its power from solar or renewable energy.[6] Local governments can create a tiered system where the more solar or renewable energy installed (measured by energy output or percentage of energy needed on-site), the higher the density bonus. These ordinances require local governments to determine the proper incentive and the corresponding property requirements. These ordinances may also require certification or review prior to receiving a certificate of occupancy.

EFFECTS

Solar energy can provide numerous benefits once installed. Such benefits include low maintenance costs and an increase in home value.[7] Solar panels can add an average of approximately 20,000 dollars to the value of a home.[8] This increase in value benefits both the developer when they sell the house and the local government when they collect property taxes.

Density bonuses, in exchange for solar energy system installation, benefit the environment by promoting clean and renewable energy. Solar energy produces significantly less greenhouse gas emissions and harmful pollutants compared to the alternative of fossil fuels.[9] By encouraging more developments to run on solar power, local governments can reduce their overall emissions.

Finally, density bonuses can help reduce energy sprawl. By providing the bonus for units within the development that provide clean energy, a local government can limit the need for an additional solar farm that would take over another parcel of land.[10] Moving towards green energy has directly affected thousands of square kilometers of land through energy sprawl.[11] Energy sprawl is the amount of energy that can be produced in a given amount of space.[12] Some governments have already begun to notice the problems that come from failing to mitigate energy sprawl caused by green energy and have begun to evaluate options to solve the problem.[13] Solar energy is one potential method to limit energy sprawl as panels can be placed on rooftops.[14]

EXAMPLES

McCall, ID

McCall provides density bonuses in the Planned Unit Development (PUD) chapter of their code.[15] The PUD chapter provides the City zoning flexibility outside of the standard zoning regulations.[16] McCall uses this flexibility to provide density bonuses for a number of public policy interests, primarily, renewable energy.[17] The City provides a 10% bonus to density, if 50% of the total energy needs of the development are provided by "solar, wind, geothermal, or [an] alternative renewable energy source."[18] This of-right bonus provides early notification of the bonus and helps encourage renewable energy.

To view the provision see McCall, ID, Code of Ordinances § 3.10.024(A) (2006).

Hinesburg, VT

Hinesburg provides density bonuses in a portion of its code relating to village growth area.[19] The additional village zoning requirements can apply to each of the City's other zoning districts including commercial, industrial, and residential districts.[20] This portion of the code provides density bonuses to developers in exchange for providing public benefits that help further the goals expressed in the town plan.[21] Notably, one of the public benefits the town provides for is renewable energy technology.[22] Developers are given the density bonus if a certain percentage of the units within the new development provide the renewable energy benefit.[23] The requirement is based on the percent of the development's long-term energy needs that are provided by renewable energy.[24] This bonus allows the City to further encourage renewable energy, and help discourage sprawl in the process.

To view the provision see Hinesburg, VT, Code of Ordinances art. 2 § 2.9 (2018).

ADDITIONAL RESOURCES

Planning for Solar Energy Briefing Papers, American Planning Association (2013), https://perma.cc/3MRV-FUDZ.

Colleen McCann Kettles, A Comprehensive Review of Solar Access Law in the United States: Suggested Standards for a Model Statute and Ordinance, Solar America Board for Codes and Standards Report (Oct. 2008), https://perma.cc/R445-NHDT.

ENDNOTES

1   Solar Energy in the United States, Office of Energy Efficiency & Renewable Energy, https://perma.cc/7QTW-YV3F (last visited June 11, 2019).
2   Mari Hernandez, *Solar Power to the People: The Rise of Rooftop Solar Among the Middle Class,* American Progress (Oct. 21, 2013), https://perma.cc/DNT7-8YM3.
3   The World Bank, Density Bonus, https://perma.cc/Q755-U5UA, (last visited June 11, 2019).
4   Incentive Strategies: Density Bonuses, Fee Waivers & Expedited Approvals, University of Florida College of Law, https://perma.cc/EG8M-VT7D (last visited June 11, 2019).
5   *Id.*
6   *See* McCall, ID, Code of Ordinances § 3.10.024(A) (2006).
7   Solar Alliance Action, 7 Reasons Why Solar is Becoming More Popular, https://perma.cc/RJD9-UZGZ (last visited June 11, 2019).
8   Ashlea Ebeling, How Much Do Solar Panels Boost Home Sale Prices?, Forbes (Aug. 1, 2011), https://perma.cc/C6XH-DCQY.

9 The Environmental and Public Health Benefits of Achieving High Penetration of Solar Energy in the United States, Office of Energy Efficiency & Renewable Energy, https://perma.cc/4W9Q-F9U2 (last visited June 12, 2019).

10 *See* Hinesburg, VT, Code of Ordinances art. 2 § 2.9 (2018).

11 Anne M. Trainor, et. al., Energy Sprawl Is the Largest Driver of Land Use Changes in United States, PLOS (Sept. 8, 2016), https://perma.cc/469B-33MU.

12 Len Calderone, *What is Energy Sprawl?*, ALTERNATE ENERGY MAG. (Dec. 4, 2018), https://perma.cc/RG9P-W8E9.

13 Council on Environmental Quality, Why Farmland and Forests are Being Developed for Electricity Production; Recommendations for Better Siting, Connecticut (Jan. 4, 2017), https://perma.cc/QM57-CSEJ.

14 *See* Calderone, *supra* note 12.

15 McCall, ID, Code of Ordinances § 3.10.01.

16 *Id.*

17 *Id.* § 3.10.024.

18 *Id.* § 3.10.024(A).

19 Hinesburg, VT. Code of Ordinances art. 2 § 2.9.

20 *Id.* art. 3 § 3.1.

21 *Id.* art. 2 § 2.9.

22 *Id.* art. 2 § 2.9(3).

23 *Id.*

24 *Id.*

ENERGY AND WATER EFFICIENCY

Brandon Hanson (author)
Jonathan Rosenbloom & Christopher Duerksen (editors)

INTRODUCTION

Lowering energy and water use can drastically reduce greenhouse gas (GHG) emissions.[1] One efficient and cost-effective way to reduce energy and water use without changing user behaviors or practices is to use energy and water efficient products.[2] Some models of air conditioners, heaters, other appliances and building products use less electricity than others. This ordinance encourages individuals and developers to switch to these more energy and water efficient products by offering a variety of incentives. Some incentives are directed at reducing developer costs or increasing developer revenue in the purchase and implementation of efficient products.[3] Example local incentives include tax credits or government grants.[4] Other incentives provide developers with building and permitting process benefits, such as expedited permitting and density bonuses.[5] These energy and water efficiency ordinances can be adapted to fit a particular local community's circumstances and can be adjusted as needed to require more energy or water efficient products as the community progresses towards cutting energy and water consumption.

Along with ordinances, many local governments offer rebate programs. Rebate programs create incentives for those looking to use more efficient products. Such incentives may be offered to residential and commercial property owners. Rebates and other financial incentives are used by many jurisdictions from local governments to the federal government. One law firm has created a list of state and federal incentives that are offered across the country.[6] Ordinances incentivizing energy and water efficient products may work in collaboration with several ordinances promoting renewable energies, thus addressing both the demand and supply side (for ordinances promoting renewable energies see *Renewable Energy with Incentives*).[7]

EFFECTS

The two most direct benefits from this ordinance are the reduction of water use and of fossil fuels burned and the associated GHGs emitted that contribute to global climate change. Switching to products that use less energy and water are quick and efficient ways to reduce GHG emissions and inefficient water usage.[8] Encouraging the use of energy and water efficient products can also have significant long-term economic benefits for those paying utility fees.[9] Energy and water efficient products lower utility costs.[10] Improving energy efficiency also easies the transition to alternative renewable energy systems (for briefs addressing renewable energy systems see Promoting Renewable Energy with Incentives and Zero Net Energy Buildings), as those systems can be smaller, decentralized, and produce less electricity to accomplish their goal of replacing fossil fuel systems.[11] Other benefits include, stabilizing energy prices as the market will be affected less by fluctuating fossil fuel prices; and lowering consumers water bills and waste water treatment costs as there is less demand for water.[12]

EXAMPLES

Minneapolis, MN

Minneapolis provides floor area ratio (FAR) premiums in the downtown district for buildings that increase their energy efficiency by a minimum of 35% compared to the Minnesota energy code. Floor area premiums are based on the total acreage of a lot, if the floor area ratio is 1.0 it allows for a total floor space equivalent to the total acreage of the lot. For example, a two-story building with each floor equivalent to 50% of the lot size would be acceptable (this concept is visualized in figure 1).[13] The Minneapolis code adds a FAR premium to buildings for energy efficiency ratings that exceed the Minnesota State code. For an increased efficiency of 35% over the State code, a developer is given an increased ratio for the building and for an increase of 45% an even greater ratio, allowing for taller buildings. If a developer complies with the 45% enhanced efficiency and creates a building in the downtown district on half the total acreage of the lot, the developer would be entitled to be build four stories higher than a similar building that does not have a qualifying energy efficiency rating. The additional floor area ratio premiums can be added with other allowances to create an even greater ratio to build higher.

To view the provision see Minneapolis, MN, Code of Ordinances § 549.220 (12) (2016).

Sarasota County, FL

Sarasota County has established an Energy Economic Zone Program that allows developers of commercial property to receive different tax rebates and other financial incentives when they comply with energy efficiency standards. Eligibility for the program can be met by receiving a specific ENERGY STAR rating (energy star is an U.S. Environmental Protection Agency (EPA) energy usage rating program), a LEED (Leadership in Energy and Environmental Design) certification, or an energy improvement strategy that will reduce energy usage by 10% as a result of efficiency measures. By providing incentives to reduce energy, developers and owners of commercial properties are more likely to use efficient appliances. The tax incentives may also help attract developers building sustainably, which may create jobs and promote the local economy.[14] The Energy Economic Zone Program also creates an incentive to update products as efficiency standards need to be met every year in order to be eligible for the various incentives.

To view the provision see Sarasota County, FL, Code of Ordinances §§ 38-300. - 38.309 (2012).

Pella, IA

A natural and common place to regulate energy use is through energy utilities. Local governments, such as Pella, often work with energy providers to offer incentives related to energy efficiency. Pella's Bright Energy Solutions (BES) offers multiple incentives to residents, such as rebates for the

> **ADDITIONAL EXAMPLES**
>
> Annapolis, MD, Code of Ordinances § 6.29.010 (2017) (providing loans to land owners for the purpose of making energy efficient improvements).
>
> Charleston, SC, Zoning Ordinance § 54-299.31 (2015) (creating a point system for developers to follow, points can be earned for various actions including the use of energy efficient products, points grant various structural incentives).
>
> Columbia, MO, Code of Ordinances §§ 27-161 — 27-169 (2011) (providing a number of financial based energy efficiency programs including grants, rebates, and loans).
>
> Boston, MA, Municipal Code § 7-2.2 (f) (2014) (creating a commission that sets regulatory energy requirements for developers and owners, the commission can set fines, and require replacements of outdated systems).

purchase of energy efficient appliances and other products. These incentives can include energy efficient washers and dryers or services to improve household systems. Available for both residential and commercial customers, the program requires owners to submit the rebate form along with proof of purchase of qualifying products. For example, if a customer purchases an Energy Star refrigerator and submits the form online the customer receives a $20-50 rebate.[15] This rebate is in addition to any energy savings that an individual will receive for using an energy efficient product.

To view the rebate program see Pella, IA, Bright Energy Solutions.

To see more on Pella's Electrical Distribution Codes see Pella, IA, Code of Ordinances §§ 111.01-111.12 (current through 2018).

ENDNOTES

1   EPA, *Local Residential Energy Efficiency*, (mar. 13, 2018) https://perma.cc/SU5P-N2CF (last visited May 18, 2018).
2   *Id.*
3   *Id.*
4   *Id.*
5   *Id.*
6   Jerome L. Garciano, *Green Tax Incentive Compendium*, Jan. 1, 2018, Robinson & Cole L.L.P., https://perma.cc/NHY4-K9F3.
7   *See id.*
8   Ming Yang & Xin Yu, *Energy Efficiency Benefits for Environment and Society*, 4-5 (2015); Intergovernmental Panel on Climate Change, *Climate Change Synthesis Report 2014*, 103.
9   EPA, *Local Energy Efficiency Benefits and Opportunities*, (Mar. 18, 2018) https://perma.cc/3Z3R-XCGT (last visited May 17, 2018).
10 *Id.*
11 *Id.*
12 Energy star, *Saving Water Helps Protect Our Nation's Water Supplies*, https://perma.cc/VH6E-YXPY (last visited May 21, 2018).
13 Bill Lindeke, *Floor Area Ratio 101: This Obscure but Useful Planning Tool Shapes the City*, Minn. Post (Oct. 3, 2016), https://perma.cc/EGL2-6H97.
14 Business Observer Staff, *Window Firm Pulls in 300K in State Incentives*, Feb. 12, 2015, https://perma.cc/AJC3-5WPF (last visited June 4, 2018).
15 Bright Energy Solutions, *Bright Energy Solutions Refrigerator Rebate Form* (2018).

GREEN ROOFING

Alec LeSher *(author)*
Jonathan Rosenbloom & Christopher Duerksen *(editors)*

INTRODUCTION

Several municipalities have found it beneficial to offer incentives for or to make it mandatory to construct and maintain "Green Roofs."[1] Developers that are required or choose to participate in such programs may be rewarded with benefits for constructing and maintaining green roofs on a variety of buildings. The incentives can provide a developer with a variety of benefits, such as expedited permitting, increased floor area ratio (FAR) or density bonuses, or tax credits to offset the costs of green roofing. Alternatively, at least one city has required green roofs as part of development and imposed penalties for developers that do not incorporate a green roof into their building.[2]

A "Green Roof" is a roof that is used to grow plant life.[3] The vegetation can be anything including grasses, wildflowers, or agricultural products. In some situations a low maintenance collection of grasses may be most appropriate,[4] while in other instances a vegetable/fruit garden can be grown to make agriculturally productive use of the area.[5] To decrease energy and stormwater management demands (discussed below), plant life should typically cover the majority of the surface area of the roof.[6] Restrictions in size and type of vegetation are often dictated by climate and building load capacities.[7] Due to the soil used in the construction of green roofs and soil water retention, green roofs are typically heavier than traditional roofs.[8] As a result, they may require additional support to ensure their proper safety and functioning.[9] They also require a membrane lining around the soil to prevent water from damaging the structure underneath.[10]

Green roof implementation can be encouraged through municipalities initiating pilot programs, providing direct or indirect financial incentives, or passing regulation.[11] Pilot programs can be accomplished through construction of green roofs on municipal buildings to market the positive benefits of green roofs.[12] Another tool municipalities may use to encourage green roofing is through the issuance of grants, tax credits or fee waivers for the construction of green roofs.[13] Alternatively, granting developers variances or additional floor area ratio or expedited permits subject to their structure having green roofs encourages green roof construction among developers.[14] Ordinances requiring green roofs for new structures in certain districts (downtown, high-traffic area) is also an effective method municipalities may utilize to promote green roofing.[15]

EFFECTS

Green roofs help turn otherwise unproductive spaces, and often costly spaces, into environmentally beneficial spaces. The majority of rooftops in a jurisdiction are often unused spaces that serve only to direct water towards gutters and absorb heat.[16] Green roofs transform these unused spaces into productive and efficient areas that mitigate stormwater run-off from impervious roof cover, improve the energy efficiency of the building, reduce heat island effect, provide open space for use by building residents and potentially the public for educational economic and aesthetic benefits.[17]

The addition of soil and vegetation to exposed surfaces, such as roofs, can significantly reduce stormwater runoff.[18] The soil and plants turn impervious surfaces into permeable surfaces, diverting water from entering the public stormwater system.[19] This is important as it saves the local jurisdiction costs in stormwater management, and also reduces greenhouse gas (GHG) emissions involved with water treatment.[20] In addition to helping with stormwater management, green roofs can help insulate the building reducing both heating and cooling costs and lowering GHG emissions.[21] Green roofs may also provide a source for numerous plant based and agricultural products that may otherwise be unavailable due to a lack of open spaces.[22]

EXAMPLES

Chicago, IL

Chicago's city ordinance 17-4-1015,[23] incentivizes the construction of green roofs on buildings in the Chicago downtown mixed used district to reduce the stress rainwater runoff places on storm water management systems. The ordinance provides that developers building in the downtown mixed-used districts may receive a Floor Area Ration (FAR) bonus when green roof is applied into the developers building. This additional footage allows developers to maximize profit by constructing more units on the same lot. The ordinance has several requirements developers must satisfy to receive the bonus. Requirements include: a green roof must cover fifty percent of the net roof area or 2,000 square feet of contagious roof area, and the roof must be structurally fit to withstand the weight of the green roof. This ordinance is notable because the city saves on stormwater management systems in the long-term and reduces the urban heat island effect plaguing Chicago's downtown in the summer months.

To view the provision see Chicago, IL, Municipal Code of Chicago § 17-4-1015 (2017).

In addition, Chicago requires that most dwelling units within the downtown area maintain a minimum amount of open space on-site.[24] The "open space must be outdoors and designed for outdoor living, recreation, or landscaping."[25] Each development is required to supply at least 36 square feet of open space for each dwelling unit.[26] This space can be supplied on the ground, as decks or patios, or on roofs.[27] The calculation to determine the amount of space necessary is to multiply the number of units for the planned development by 36 square feet.[28] So a 100 unit development would need to supply 3,600 square feet of open space in total. One option to fulfill this requirement is by using some combination patios and ground level open space. However, in areas where ground space is in high demand, it may be more efficient to locate the required open space on roofs.

To view the provision see Chicago, IL, Municipal Code of Chicago § 17-4-0410 (2017).

Denver, CO

Denver's Denver Green Roof ordinances requires the construction of green roofs for new buildings over 25,000 square feet, and for existing buildings at least 25,000 square feet when the building requires a roof replacement. The code defines a green roof as an extension above the roof, which allows vegetation to grow. The portion of the roof required to be dedicated to green roofing increases along with the gross square footage of the building. For example, a 25,000 square foot building must be covered by 20% green roof, whereas a building with 200,000 square foot or more must be covered by 60% green roof. Owners may combine solar paneling and green roofing to meet the requirements. The ordinance further creates a green roof technical advisory board, which is composed of experts in green roofing and advises the city planning board on green roof issues. The planning board and technical advisory board also promulgate guidelines for the construction of green roofs, to assist developers in complying with the ordinance.

The ordinance imposes penalties for property owners that forego construction of green roofs for buildings that received a permit to build after January 1, 2018.[29] If a property does not comply with the requirements, the owners or occupiers thereof are subject to a fine of no more than $999 per violation, or imprisonment for a maximum of one year per violation. A violation occurs every 24 hours that the property is out of compliance with green roof requirements. Voters approved these measures by referendum, hoping to reduce urban heat island effects, provide habitat for bees and other wildlife, and reduce GHG emissions. Notably, the requirements are not imposed on residential buildings less than 4 stories or 50 feet in height, or greenhouses and their related structures.

To view the provision see Denver, CO, Code of Ordinances §§ 10-300 to 10-308 (2017).

ADDITIONAL EXAMPLES

Philadelphia, PA, Philadelphia Code § 19-2604 (8) (2017) (providing for tax credits available for qualifying green roofs).

Austin, TX, Land Development Code § 25-2-586 (E) (11) (2014) (creating a bonus floor area incentive program).

Minneapolis, MN, Zoning Code § 527.120 (2009) (adding green roofs to a program of alternatives to zoning standards, if a developer desires to adjust some zoning regulations they may do so by adding amenities one of which is a green roof).

San Francisco, CA, Planning Code § 149 (2016) (allowing "living roofs" to meet the requirements of a state law that mandates a portion of roofs be "solar ready").

Portland, OR, City Code § 33.510.210 (C) (5) (2018) (granting an extra square foot of floor area ratio for every square foot of green roof incorporated onto the building).

Nashville, TN, Code of Ordinances § 15.44.050 (E)(b) (2012) (granting a $10 rebate on a private property owner's sewer bill for each square foot of qualifying green roof installed).

Sustainable Land Use Code Project § 1.2.2 (Capital Region Council of Gov'ts 2013) (model green roof ordinances that provides floor area bonuses, open space credit, and stormwater management credit).

New York City, NY

New York City's Green Roof Property Tax Abatement Program[30] is a direct incentive tool utilized by New York City to encourage retrofitting of green roofs or construction of green roofs on new structures. The program offers building owners or developers $4.50 per square feet of green roof or a maximum abatement of $100,000 or the building tax liability (whichever is lesser) subject to several building and compliance conditions. Some of these conditions include: (1) construction of a green roof must have commenced after August 5, 2008; (2) at least fifty percent of eligible roof space must be covered by the green roof: (3) guidelines on structural compliance of the green roof; and (4) post construction maintenance. Due to the congestion in New York City and lack of green space, this program hopes to encourage construction of green roofs, that will help reduce the urban heat island effect. This program does not place a significant burden on the city's budget, because the resources used to monitor this program currently exists—architects and engineers employed by the Department of Building. Additionally, this program has significant economical benefits to developers and property owners, and provides for a healthier and cleaner environment.

To review the provision see New York City, NY, Zoning Code 1 RCNY §105-01 (2008).

ENDNOTES

1   Chicago, IL, Municipal Code of Chicago § 17-4-1015 (2017); Portland, OR, City Code § 33.510.210 (C) (5) (2018); Minneapolis, MN, Zoning Code § 527.120 (2018); Denver, CO, Code of Ordinances § 10-301 (2018).
2   Denver, CO, Code of Ordinances § 308(b)(2) (2018).
3   Emily W, O'Keefe et al., *Raise the Roof: Green roofing options offer lower energy costs and better aesthetics*, 2008 J Prop. Mgmt. 64, 64; David Johnston & Kim Master, Green Remodeling: Changing the World One Room at a Time, 212-213 (2004); US Environmental Protection Agency, *Soak Up the Rain: Green Roofs* https://perma.cc/VFS6-4AZP (last visited May 14, 2018).
4   Dyanna Innes Smith, Green Technology: An A-to-Z Guide, *Green Roofing*, 230, 231-32 (Dustin Mulvaney ed., 2011); Johnston & Master, *supra* note 3, at 213.
5   Smith, *supra* note 4, at 231-32; Johnston & Master, *supra* note 3, at 213.
6   O'keefe, *supra* note 3, at 64
7   O'keefe, *supra* note 3, at 64
8   Johnston & Master, *supra* note 3, at 213.
9   *Id.*
10   Johnston & Master, *supra* note 3, at 212-213; Dyanna Innes Smith, *supra* note 4, at 231.
11   *See* Catherine Malina, *Up on the Roof: Implementing Local Government Policies to Promote and Achieve the Environmental, Social, and Economic Benefits of Green Roof Technology*, 23 Geo. Int'l Envtl. L. Rev. 437 (2011).
12   City of Chi. Dep't of Env't, Green Roof Test Plot 2003 End-of-Year Project Summary Report (2004), http://egov.cityofchicago.org/webportal/COCWebPortal/COC_ATTACH/2003GreenRoof-Report.pdf.
13   New York City, NY, Zoning Code 1 RCNY §105-01 (2018).
14   Chicago, IL, Municipal Code of Chicago § 17-4-1015 (2017).
15   Denver, CO, Code of Ordinances § 10-301 (2018) (newly enacted legislation in Denver, Colorado providing for buildings constructed after January 1, 2018 with 25,000 square feet to have a certain percentage of roof space covered by green roof or solar panel.); Jackie Snow, Green Roofs Take Root Around the World (October 27, 2016), https://perma.cc/3YVA-C8KD.
16   Richard K Sutton, *Introduction to Green Roof Ecosystems*, 223 Ecological Studies 1, 3-5 (2015).
17   *See* Malina, *supra* note 11; Environmental Protection Agency, Using Green Roofs to Reduce Heat Islands, https://perma.cc/6SDL-W8M5.

18 O'Keefe, *supra* note 3, at 66; Johnston & Master, *supra* note 3, at 212-13.
19 Johnston & Master, *supra* note 3, at 212-213; Sutton, *supra* note 16, at 3-5.
20 Sutton, *supra* note 16, at 5.
21 Smith, *supra* note 4, at 233; O'Keefe, *supra* note 3, at 66; Johnston & Master, *supra* note 3, at 212-13.
22 O'Keefe, *supra* note 3, at 66.
23 Chicago, IL, Municipal Code of Chicago § 17-4-1015 (2017).
24 *Id.* § 17-4-0410-A.
25 *Id.* § 17-4-0410-B (2)
26 *Id.* § 17-4-0410-A.
27 *Id.* § 17-4-0410-B (2).
28 *Id.* § 17-4-0410-A.
29 Denver CO, Code of Ordinances § 308(b)(2) (2018).
30 New York City, NY, Zoning Code 1 RCNY §105-01 (2018).

INFILL DEVELOPMENT

Tyler Adams (author)
Jonathan Rosenbloom & Christopher Duerksen (editors)

INTRODUCTION

Infill development is the process of developing vacant or under-developed parcels within areas that are already largely developed.[1] As populations fluctuate and the needs of a community transform, vacant land becomes increasingly common place. Instead of directing development outward, infill development helps replace existing vacant lots and promotes land conservation through the reduction of greenfield development. Successful infill development programs often focus on improving neighborhoods, creating more efficient mixes of jobs and housing, reducing blight, and reinvesting in the community.[2] Infill can return cultural, social, and recreational vitality to dilapidated areas within a community.[3]

Infill development occurs on a variety of scales, including the rehabilitation of an entire block or the construction of a single-family home on a vacant lot within a developed block.[4] Different infill programs may be appropriate depending on the type of development needing to be achieved. Infill developments are also not limited to a particular type of use. Municipalities frequently permit residential, commercial, or a mixture of uses in infill development areas in order to accomplish a variety of goals. Traditional barriers to infill development include neighborhood opposition, inflexible building codes and difficulty in land assembly.[5] To address some of these barriers many communities have created incentive programs to make infill a more attractive option. These incentives may reduce fees or relax building and design requirements. When creating incentives to develop infill areas, local governments may attach a variety of sustainable building requirements, such as minimum energy production (see Zero Net Energy Buildings.

EFFECTS

Encouraging infill development can positively impact a community in a variety of ways. Vacant and underdeveloped lots are typically integrated, or readily able to be integrated, into existing infrastructure, including sewers, roads, and public transit services.[6] This greatly reduces the cost of development and the need for additional resources associated with having to connect to these essential utilities. Further, infill development helps combat sprawl, which is often comprised of low density development and the separation of uses, thus increasing a community's reliance on automobiles.[7] Infill development can increase the density of an area which, in combination with expanded public transit, can decrease the emission of greenhouse gases.[8] Developing homes in close proximity to existing public transit and integrating non-residential and residential uses reduces the number of vehicle miles traveled.[9] By reducing sprawl, infill development also conserves natural resources, protects biodiversity, and promotes watershed protection.[10] A community is also able to benefit from the redevelopment and cleanup of areas that may be contaminated with hazardous substances or pollutants.[11]

By facilitating the concentration of development in established areas and on dilapidated sites, a municipality can provide businesses with an already established market.[12] The businesses and the municipality are able to economically benefit from the increased spending within city limits.[13] Furthermore, vacant and abandoned lots cost a municipality money and pose numerous safety hazards, such as increased risk of fires and crime. Each year over 12,000 fires break out in vacant buildings in the U.S., most of which are the product of arson, resulting in $73 million in property damage annually.[14] Municipalities frequently will also have to bear the cleaning and demolition cost associated with these properties.[15] Additionally, neighboring property owners are impacted by vacant lots through decreased property values of their properties.[16] Redevelopment of these areas is not only essential to restoring the economic viability of a neighborhood, but it also permits the preservation of historical areas through their revitalization.[17]

EXAMPLES

Avondale, AZ

Avondale established an infill incentive program aimed at certain neighborhoods in order to reinvigorate existing historic areas and support new mixed-use development that would promote the historic identity of the area.[18] The neighborhoods targeted by the infill incentive program contain many vacant or underutilized areas and exhibit at least one of the following characteristics: high vacancy rates, larger number of older buildings, and continued decline in population in relation to the City as a whole.[19] Qualified residential projects on residentially-zoned property within the infill incentive district are subject to reduced fees. New residential construction projects receive a 50% reduction in the normal planning and permit fee as well as the development impact fee.[20] Rehabilitation or remodeling projects are allowed

a 50% reduction in planning and permit fees and are not subject to development impact fees.[21] Qualified commercial projects have the same incentives as residential projects, but with the added benefit of additional incentives being available subject to approval by the city council.[22] These additional incentives are based on additional criteria, such as high-wage job creation.[23]

To view the provision see Avondale, AZ, Code of Ordinances § 19-61 (2014).

Aurora, CO

A developer intending to develop or redevelop a parcel contained in the "infill incentive boundary area" is able to apply for an incentive under this ordinance.[24] To be eligible for the program, the project must comply with certain criteria. For example, commercial infill developments must be limited to 5,000 square feet or less for a single-story project or 10,000 square feet or less for a multi-story project. Further, residential infill projects are limited to a minimum of two residential units and a maximum of eight units.[25]

Once the criteria are met, the developer and the City are able to enter into an agreement that provides for a grant to offset development related fees as well as sale and use taxes. For instance, for a period of up to two years after the agreement, the City may reimburse the developer, contractor, subcontractor, or supplier for the materials and equipment used during construction by up to 50% of the revenues produced by the levy of the City sales tax.[26] The City also places a cap of $25,000 on the amount that can be offset through the incentive program.[27]

To view the provision see Aurora, CO, Code of Ordinances § 130-564 (2016).

ENDNOTES

1   *Infill Development*, MRSC Local Government Success (Nov. 30, 2017), https://perma.cc/XYS5-VUPS.
2   *Infill Development*, Connect Our Future, https://perma.cc/7G69-4CBA (last visited May 21, 2018).
3   MRSC, *supra* note 1.
4   *Id.*
5   *Id.*
6   Robert H. Freilich, *Smart Growth in Western Metro Areas*, 43 Nat. Resources J. 687, 697 (2003).
7   *See* Michael P. Johnson, *Environmental Impacts of Urban Sprawl: A Survey of the Literature and Proposed Research Agenda*, 33 Env't and Plan. A: Econ. and Space 717, 718 (2001).
8   Sarah DeWeerdt, *Urban Density Alone Won't Get Americans Out of Their Cars*, Anthropocene Magazine (Dec. 26, 2017), https://perma.cc/9CGV-3RPV.
9   *Id.*
10   *See* University of Delaware, *Benefits of Infill and Redevelopment Activities*, Complete Communities, https://perma.cc/634J-JS86 (last visited May 23, 2018).
11   *Id.*
12   Kelli Russel & Kelsey Knight, *Urban Infill & Sustainable Development* 5 (2013).
13   *Id.*
14   National Vacant Properties Campaign, *Vacant Properties: The True Cost to Communities*, Smart Growth America (Aug. 2005), https://perma.cc/L6MS-9ZXC.
15   *Id.*
16   *Id.*
17   University of Delaware, *supra* note 10.
18   Avondale, Arizona Code of Ordinances § 19-61 (2014).
19   *Id.*
20   *Id.* §19-64.
21   *Id.*
22   *Id.* § 19-65.
23   *Id.*
24   Aurora, Colorado Code of Ordinances § 130-564 (2016).
25   *Id.*
26   *Id.* § 130-565.
27   *Id.*

LIMITING OFF PROPERTY SHADING OF SOLAR ENERGY SYSTEMS

Trisana Spence, Kathryn Leidahl (authors)
Darcie White, Sara Bronin, & Jonathan Rosenbloom (editors)

INTRODUCTION

Renewable solar energy systems create no direct pollution in their functioning and require very little maintenance.[1] As a result, solar energy has become a popular way to generate electricity, provide light, and heat water for domestic, commercial, and/or industrial use.[2] To operate efficiently, solar energy systems need direct access to sunlight. The more direct sunlight, the more efficient the system is. Shading from vegetation and structures can block sunlight. This presents a difficult challenge when the vegetation and structures are on someone else's property. To balance the necessity of new vegetative and developmental growth against the importance of solar access rights, some local governments have enacted legislation that expressly protects solar energy systems from shading.

Protection from shade can be achieved through a permit process, by-right, or through an easement.[3] In terms of permits, there are two significant permit processes that create protection from shading: a protective solar access permit and a solar installation permit. The two achieve the same goals, but in slightly different ways. A protective solar access permit ensures the shading from vegetation, fences, buildings, and other structures on adjacent properties do not interfere with solar energy collectors. A solar installation permit grants the installer/owner of the solar energy system protected access. For more information regarding solar installation permits see the Streamlining Solar Permit and Inspection Process brief.

The language included in local permitting regulations may include limiting the height of existing trees and buildings, ensuring new construction will not interfere with potential or existing solar energy systems, and prohibiting the planting of certain vegetation if when fully grown would shade or interfere with solar energy system. Obtaining a permit may include an application process, which requires submission of a map of all properties affected, labeling buildings, vegetation, and other structures. After the application is complete, the burdened landowner may be put on notice that an application has been submitted and afforded an opportunity to raise objections to the permit. An individual or committee appointed by the local government will then have the ability to either approve or deny the application. If approved, the landowner will receive either a protective permit over the system or an installation permit to build a system. The landowner typically has the burden of recording the permit and installing or maintaining the solar energy system while the burdened landowner has the responsibility of removing and maintaining vegetation and structures that have the potential to interfere with the system.[4]

EFFECTS

Creating an opportunity for landowners to protect their solar rights can incentivize solar energy systems by protecting the efficiency of the systems. Prohibiting new vegetative growth and other structures from shading solar energy systems protects solar investments. In order to gain a quicker return on initial investment, a system needs to be clear of shading to operate at full capacity. When a solar energy system underperforms because of shading outside the owner's control, owners anticipated return on investment is extended, adding financial hardship and increasing greenhouse gases. Ordinances which grant protection from shading reduce the risk of underperforming systems by legally obligating the surrounding landowners to eliminate structures and vegetation that obstruct direct sunlight.

Drawbacks to enacting a protective ordinance prohibiting new vegetative growth and structures from shading include that it may impose unnecessary or unexpected economic costs on neighbors in both maintenance and property value. By limiting development rights, the ordinance may alter property values in a manner similar to other regulatory use impacts on property and may increase permitting burdens.[5]

EXAMPLES

Hartford, CT

Hartford, CT enacted a provision that prohibits new vegetative growth from shading solar energy systems. Because this ordinance establishes the protection "by-right," no legislative or administrative action or protective permit is necessary to protect solar energy systems. The system is protected by-right from new vegetative growth and associated shading. The regulation states that "a property owner may not plant any tree which, when fully grown, will shade a solar collector existing at the time of the planting of the tree."[6]

To view the provision see Hartford, CT, Zoning Regulations §6.4.1(E) (2018).

Ashland, OR

In 1981, Ashland, OR, was one of the first cities in the country to introduce a solar access protection law.[7] The primary purpose of the City's solar access ordinance is to preserve the investment in solar energy by decreasing the amount of shade from structures and vegetation whenever possible.[8] Any property owner may apply for a protective permit as long as their application includes a $50 fee with an additional $10 fee for each lot affected by the permit, a statement by the applicant, affirming "the solar energy system is already installed or that it will be installed on the property within one year following the granting of the permit," and a parcel map of the owner's property, identifying the location of adjacent buildings and vegetation.[9] A Staff Advisor reviews the permit application, allows 30 days for any objections by adjacent landowners, then notifies the applicant if the permit has been approved. Once the permit is in place, planting vegetation that shades the system or erecting a structure that interferes with the system is prohibited.[10] Further, in Ashland, a Solar Access permit "becomes void if the use of the solar energy system is discontinued for more than 12 consecutive months or if the solar energy system is not installed and operative within 12 months of the filing date of the solar access permit."[11]

To view the provision see City of Ashland, OR, Land Use Code §18.4.8.060 (current through December, 2018).

Ridgecrest, CA

The City of Ridgecrest, CA creates a solar access easement to prevent vegetation, buildings, fences, walls, or other structures from creating shade over a solar energy system.[12] Located in the subdivision code, the request for an easement must contain a clear diagram of the direct sunlight path and language for the steps to terminate or revise the easement.[13] If the easement drastically alters the purpose of the burdened land or would not be feasible, the structures or vegetation that cast a shadow will remain and the easement will not be valid.[14] Additionally, if the easement is approved without any issues, it should be recorded to be valid.[15]

To view the provision see Ridgecrest, CA, Code of Ordinances § 19-2.4 (current through 2017).

Boulder, CO

Boulder, CO, divides properties into three "Areas" to ensure each type of property receives adequate solar protection.[16] Area One, which contains multiple zones, primarily protects south yards, walls, and rooftops.[17] There is no permit required in this area to receive protection.[18] Area Two only protects rooftops and also does not require a permit to gain protection.[19] Area Three, which consists of planned densities, requires a permit to gain solar access protection.[20]

The City's ordinance calculates the applicable degree of shade protection by the height of a vertical and hypothetical "solar fence." Such a fence completely encloses the lot within the property lines.[21] In Area One, a structure may not be built which would shade the lot greater than a hypothetical twelve-foot "solar fence."[22] In Area Two, the hypothetical "solar fence" height is increased to twenty-five feet.[23] The solar fence calculation is not applicable in Area Three, as the only way to obtain protection in this area is through a protective permit.[24]

When an individual in the City requires a permit to obtain solar access protection, their application must include a description of the solar energy system, the adjacent properties, any obstructing vegetation, the benefit from the system, and proof that the system is or will be located in an area where its efficiency is maximized.[25] Once all the application materials have been received, the City Manager has the authority to approve the permit and may place additional conditions on the permit.[26] If the permit is approved, it will expire if the system is not installed within one year or is dormant for a continuous period of two

---

ADDITIONAL EXAMPLES

Clackamas, OR, Municipal Code § 1019.01 (2005) (providing for a protective permit that prohibits shade caused by vegetation and adjacent land structures).

Schaumburg, IL, Municipal Code § 154.70 (2016) (providing for protective easements against shade by vegetation or structures, such easements may be granted, reserved, bought, or otherwise obtained).

South Miami, FL, Land Development Code § 20-3.6(w)(3) (Current through 2018) (requiring many residential structures to be designed in a way which maximizes the efficiency of existing and future solar collectors).

years.[27] Additionally, the protection permit is only enforceable if it has been properly recorded.[28]

In regards to vegetation, any greenery growing or in the ground at the time the permit application is submitted may not be ordered trimmed or removed.[29] If future vegetation interferes with the solar collector after the permit has been granted, the cost to remedy the shading falls on the vegetation owner.[30] If the individual fails to remedy the problem, the City will complete the project at the individual's expense.[31] Additionally, failure to pay the City for the removal may result in a collection by the County Treasurer and may be levied against the individual's personal property.[32]

To view the provision see City of Boulder, CO, Municipal Code § 9-9-17(c).

ENDNOTES

1   The Advantages of Solar Power, Alternative Energy, https://perma.cc/2CEJ-98CS (last visited June 18, 2019).
2   About Solar Energy, Solar Energy Industries Association, https://perma.cc/8JVD-BR6T (last visited June 18, 2019).
3   Colleen McCann Kettle, A Comprehensive Review of Solar Access Law in the United States, Solar America Board for Codes and Standards, (Oct. 2008) https://perma.cc/9ZBX-JQ3U.
4   City of Boulder, CO, Municipal Code § 9-9-17(c) (2012).
5   Protecting Solar Access, SF Environment, (Dec. 2018) https://perma.cc/2V4Q-36AK.
6   Hartford, CT, Zoning Reg. §6.4.1(E) (2018).
7   Solar Powering Your Community, University of Nevada, Las Vegas, (Dec. 7, 2018) https://perma.cc/U7TA-SA2S.
8   City of Ashland, OR, Land Use Code §18.4.8.060 (current through December 2018).
9   Id.
10 Id.
11 Id.
12 Ridgecrest, CA, Code of Ordinances § 19-2.4 (2017).
13 Id. § 19-2.4 (b)(1-4).
14 Id. § 19-2.4(c).
15 Id. § 19-2.4(d).
16 City of Boulder, CO, Municipal Code § 9-9-17(c) (2018).
17 Id.
18 Id.
19 Id.
20 Id.
21 Id. § 9-9-17(d)(1).
22 Id. § 9-9-17(d)(1)(A).
23 Id. § 9-9-17(d)(1)(B).
24 Id. § 9-9-17(d)(1)(C).
25 Id. § 9-9-17(h)(3)(A-I).
26 Id. § 9-9-17(h)(5).
27 Id. § 9-9-17(h)(10)(A-C).
28 Id. § 9-9-17(h)(13).
29 Id. § 9-9-17(h)(14)
30 Id.
31 Id.
32 Id.

PERVIOUS COVER MINIMUMS AND INCENTIVES

Kerrigan Owens (author)
Jonathan Rosenbloom & Christopher Duerksen (editors)

INTRODUCTION

This regulation provides local governments with the flexibility to either create incentives, set requirements, or do a combination of both for the minimum use of permeable pavements for certain projects. Local governments may choose the type and size of projects to be subject to permeable requirements or incentives and may choose the appropriate level of permeable surfaces.

Permeable pavement or surfaces refer to any paving system that provides a usable hard surface but also allows for infiltration of water through the surface.[1] Permeable pavements come in a number of different forms, with new technologies and techniques constantly being developed.[2] Such systems include a variety of surfaces, such as pavers that connect to form a surface with small gaps to allow water to pass through[3] and porous concrete that forms a single or multi-slab coarse concrete surface. Porous concrete has a number of small gaps throughout to allow water to flow though the surface.[4] The different types of permeable pavement have unique costs and benefits that developers should consider and that local governments should consider when drafting this ordinance.[5] Some alternatives may be less desirable for high-traffic areas such as highways or areas with significant snow and freezing temperatures that limit permeability.[6]

EFFECTS

Traditional concrete pavement, asphalt, and other impermeable surfaces have a number of potentially adverse environmental effects.[7] While most codes may require some type of paved surfaces, there are several alternatives that can replace or reduce the detrimental effects of impermeable pavement.[8] Doing so may help divert run-off from entering into local stormwater management systems or a bodies of water.[9] Local stormwater utilities that often maintain storm sewers and other drainage systems bear the costs associated with impermeable pavement.[10] In addition, run-off can increase flooding either in the municipality or downstream.[11] This run-off may add pollutants to the water requiring additional costs and energy to be spent to remove them.[12] Water treatment is an energy intensive process demanding use of fossil fuels that can result in significant greenhouse gas (GHG) emissions.[13] In addition, reducing the amount of storm water and associated pollutants entering the system can lower water treatment costs.

EXAMPLES

Los Angeles County, CA

Los Angeles County encourages the use of permeable pavement through Low-Impact Development (LID) standards. This ordinance specifically addresses the Santa Monica Mountains local implementation coastal development plan that is working to protect and manage the areas resources.[14] Some of the requirements include minimizing impervious surfaces such as sidewalks, driveways, or parking, using permeable materials when possible, and directing new impervious surface run-off to permeable areas.[15] The relevant provisions provide that a development will be given "preferential consideration for approvals" if using techniques that minimize impacts due to run-off.[16] Several recommended techniques are then set forth in the code including: the reduction of impervious surfaces, the redirection of run-off into permeable areas, and the prioritization of permeable pavement over traditional pavements.[17] A project following these guidelines may reduce run-off and improve water quality.[18]

To view the provision see Los Angeles County, CA, Code of Ordinances § 22.44.1340 (G).

San Antonio, TX

As part of its Low Impact Development and Natural Channel Design Protocol (LID/NCDP), San Antonio encourages the use of permeable pavement by providing both permitting credits as well as a stormwater fee discounts for landscaping, parkland, tree canopy, buffering, and storm water to developers using the LID/NCDP.[19] However, to qualify for this credit the development must be able to manage at least 60% of the stormwater run-off that the development will generate.[20] To meet the 60% goal the LID/NCDP provides a number of best practices that should be used by the developer, including the use of permeable pavement.[21] Permeable pavement is allowed for both parking and sidewalks,[22] with a specific recommendation for the use of permeable pavement for the construction of any parking spaces above the required minimum.[23] This portion of the ordinance allows the developer to increase the amount of land dedicated to parking, but does so in a way that will not increase to amount of run-off generated by those additional parking spaces. If a developer using these best management practices reaches the 60% goal, s/he is awarded a 5% discount on stormwater management fees.[24] This discount also scales up if a developer is able to meet a higher percentage of the stormwater volume that can be managed--up to a 30% discount.[25]

To view the provision see Sec. 35-210 Low Impact Development and Natural Channel Design Protocol (LID/NCDP).

Fairway, KS

Fairway has enacted ordinances that set mandatory permeable surface minimums for new development within the city.[26] The ordinance mandates a percentage of permeable and open space for Single Family Residential Districts, Business Districts, and Mixed Use Districts. For example, within the Single Family Residential Districts, any lot under 10,000 square feet must meet the 60% permeability standard.[27] This regulation also applies to lot sizes between 10,000 square feet and 30,000 in which the first 10,000 square feet must meet the 60% permeable requirement, and the remaining lot must meet 75% permeable requirement.[28] Finally, for lots over 30,000 square feet, the first 10,000 square feet must be 60% permeable, up to 30,000 square feet must meet the 75% permeability rate, and the remaining square footage must be 100% permeable.[29] The city also requires a mandatory minimum of green space within Mixed Space districts, and requires specific permeability and vegetation minimum for the space.[30]

To view the provision see Fairway, KS, Code of Ordinances Sec. 15-264 Zoning Districts.

Minneapolis, MN

Minneapolis enacted an ordinance in 2010 to revise its zoning code to allow pervious pavement for driveways.[31] Prior to 2010, most of the pervious pavement for driveways was not permitted or required a variance.[32] Under this ordinance, pervious pavement is permitted for driveways in all residential, commercial, and industrial districts subject to certain conditions and restrictions under the applicable provisions of the zoning chapter.[33] Namely, the ordinance requires that the pervious pavement or pervious pavement system be capable of carrying a wheel load of 4,000 pounds and installed per industry standards.[34] The ordinance also prohibits pervious pavement in areas used for the distribution of gasoline or other hazardous liquids that could be absorbed into the soil through the permeable pavement.[35] These conditions and restrictions ensure that the pervious pavement meets adequate safety and performance standards. The conditions and restrictions also protect the environment by ensuring that the soil is not exposed to hazardous liquids. In short, Minneapolis's zoning ordinance for pervious pavement demonstrates a municipality's solution to managing urban stormwater challenges by allowing permeable pavement for driveways.

To view this provision see Minneapolis, MN, Zoning Code § 541.305.

ENDNOTES

1   Vermont Department of Environmental Conservation, *Pervious Pavement*, 1 (2017).
2   Benjamin O. Brattebo & Derek B. Booth, *Long-term Stormwater Quantity and Quality Performance of Permeable Pavement Systems*, 37 Water Research 4369, 4371 (2003); Vermont Department of Environmental Conservation, *supra* note 1.
3   Brattebo & Booth, *supra* note 2, at 4371; Vermont Department of Environmental Conservation, *supra* note 1.
4   Brattebo & Booth, *supra* note 2, at 4371; Vermont Department of Environmental Conservation, *supra* note 1.
5   Brattebo & Booth, *supra* note 2, at 4371; Vermont Department of Environmental Conservation, *supra* note 1.

6   Brattebo & Booth, *supra* note 2, at 4371; For additional information and photos see Vermont Department of Environmental Conservation, *supra* note 1.
7   Brattebo & Booth, *supra* note 2, at 4369.
8   ANDREW KAVARVONEN, POLITICS OF URBAN RUNOFF: NATURE, TECHNOLOGY AND THE SUSTAINABLE CITY 11 (2011).
9   AN LIU ET AL., ROLE OF RAINFALL AND CATCHMENT CHARACTERISTICS ON URBAN STORMWATER QUALITY 2-3 (2015); KAVARVONEN, *supra* note 8, at 11.
10   KAVARVONEN, *supra* note 8, at 18.
11   AN LIU, *supra* note 9, at 3.
12   *Id.* at 4-5; KAVARVONEN, *supra* note 8, at 13.
13   Environmental Protection Agency, *Energy Efficiency in Water and Wastewater Facilities,* 1 (2013) https://perma.cc/3UYJ-5XX7.
14   Los Angeles County, CA, Code of Ordinances § 22.44.1512.
15   *Id.* at § 22.44.1340 (G).
16   *Id.*
17   *Id.*
18   *Id.*
19   San Antonio, TX, Code of Ordinances §§ 35-210 (b) (2) (A), (B).
20   *Id.* § 35-210 (b) (2) (B) (2).
21   *Id.* § 35-210 (j) (3).
22   *Id.* §§ 35-210 (f) (5), (6).
23   *Id.* § 35-210 (j) (1).
24   *Id.* § 35-210 (b) (2) (B) (2).
25   *Id.* Table 210-2
26   Fairway, KS, Code of Ordinances §§15-264, 15-456.
27   *Id.* § 15-297.
28   *Id.*
29   *Id.*
30   *Id.* § 15-407.
31   *See* Minneapolis, MN, Zoning Code § 541.305 (2017).
32   CITY OF MINNEAPOLIS, COMMUNITY PLANNING AND ECONOMIC DEVELOPMENT PLANNING DIVISION REPORT ZONING CODE TEXT AMENDMENT 2 (2010), https://perma.cc/3MK5-US72.
33   Minneapolis, MN, Zoning Code §§ 541.300, 541.305 (2017).
34   *Id.* § 541.305(a).
35   *Id.* § 541.305(a)(4).

PRIORITY PARKING FOR HYBRID & ELECTRIC VEHICLES

Kyler Massner (author)
Jonathan Rosenbloom & Christopher Duerksen (editors)

INTRODUCTION

Local governments can remove barriers, create incentives, or make requirements that encourage the adoption of hybrid electric vehicles (HEVs) and electric vehicles (EVs). Barriers preventing wide-scale adoption of HEV and EV technology include inefficient permitting processes and inaccurate categorization of uses within the development code. Incentives such as free or reduced parking, or requirements such as priority parking minimums, can encourage the adoption of HEVs or EVs. In addition, such programs may create public awareness that works to educate and change purchasing behaviors.

The lack of available EV infrastructure (i.e. charging stations) is a contributing factor in the rate of HEV and EV adoption. The lack of EV infrastructure is partly a result of barriers embedded within the development code.[1] One such barrier is the improper categorization of EV infrastructure.[2] Many local ordinances place electric charging stations in the same category as traditional gas stations. This categorization requires electric charging stations to have the same facilities and safety devices of traditional gas stations. In addition, this categorization prohibits them from many zoning districts which they would be most often used.[3] By accurately categorizing and defining EV infrastructure as different than traditional gas stations, local governments can remove barriers that impede investment and expand the availability of EV infrastructure.

Local governments may also consider creating incentives by establishing priority parking programs or requiring minimum priority parking stalls that encourage the adoption of HEVs and EVs. Priority parking programs are flexible and able to take advantage of existing parking infrastructure with little to no additional costs for local government.[4] Incentives can be either non-monetary or monetary.[5] Non-monetary incentives are benefits such as parking near entrances, while monetary incentives include discounted or free parking rates for owners of HEVs and EVs.[6] Such regulations can stipulate how much parking must be set aside for use by HEVs and EVs and/or how many electric charging stations are required to be installed.[7]

EFFECTS

In 2015, the U.S. transportation sector was the second largest emitter of greenhouse gases, representing 27% of total U.S. emissions.[8] Incentivizing the transition to HEVs and EVs provides an opportunity to move toward long-term reductions in greenhouse gas emissions. It also facilitates the reduction of other forms of air pollution and reduces reliance on foreign oil.[9] Electrifying the transportation sector can drastically reduce smog and harmful ground level ozone, and thus provide a variety of health benefits.[10]

Both state and federal governments have passed laws and regulations that decrease costs and create incentives for the adoption of HEVs and EVs.[11] To achieve the highest adoption rates of HEVs and EVs, the best practice is to offer multiple, simultaneous, and parallel incentive structures such as investing in EV infrastructure, diversifying financing options, making more robust purchasing incentives, and offering other non-monetary benefits (e.g. priority parking).[12] States that had the most incentives had approximately two to four times the national average in sales of HEVs and EVs.[13]

EXAMPLES

Chelan, WA

The City of Chelan, WA encourages the use of EVs by eliminating barriers that impede installation of electric charging infrastructure. First, the City defines what an EV charging station is and then places it into three categories. This clarification encourages installation of EV infrastructure by removing barriers in the development code and providing construction standards.[14] By doing such, the City clearly establishes where charging stations can be located, and eliminates unnecessary zoning barriers that would have kept smaller charging stations from residential zones.[15] As a result, smaller charging stations are allowed in all zoning districts, while larger stations are either permitted or are granted approval upon a conditional permit in commercial or industrial zones. Furthermore, such clarification no longer requires charging stations to have the same safety infrastructure as traditional gas stations.[16]

The City also incentivizes the use of EV charging stations by allowing developers to include charging stations in the calculation for minimum parking requirements. A common disincentive to the construction of charging infrastructure is how it relates to minimum parking requirements. Often EV charging stations do not count toward minimum parking requirements. This unnecessarily requires developers to add extra conventional parking spaces at a significantly increased costs. The City of Chelan removes this disincentive by permitting developers to include an electric charging station within the calculations for the minimum required parking spaces.

To view the provision see City of Chelan, WA, Municipal Code § 17.63 (2018).

Atlanta, GA

The City of Atlanta recognizes the importance of EVs and HEVs and their ability to reduce harmful emissions, improve air quality, and further the City's commitment to increasing local sustainability.[17] The City facilitates the adoption of EVs and HEVs by investing in EV infrastructure and removing barriers within the City's codes.[18] First, the City accurately defines EV infrastructure and distinguishes it from traditional gas stations. The ordinance describes the different types of charging stations and provides regulations for their use. The City also removes a regulatory obstacle that would have converted surface parking areas into a "service station" because of the presence of an electric charging station. Such clarification removes unnecessary burdens and increases the availability and opportunities to construct EV infrastructure.

To view the provision see City of Atlanta, GA, Code of Ordinances § 16-29.0019(56) (2014); City of Atlanta, GA, Ordinance No. 2014-53 (14-01278) (May 12, 2014).

ADDITIONAL RESOURCES

Office of Transportation and Air Quality, *Fast Facts: U.S. Transportation Sector Greenhouse Gas Emission 1990-2015*, EPA (July 2017), https://perma.cc/PTC8-GMWZ (last visited June 19, 2018).

Office of Energy Efficiency and Renewable Energy, *Electric-Drive Vehicles*, DOE (Sept. 2017), https://perma.cc/R3AX-XAM9 (last visited June 19, 2018).

Yan Zhou, Todd Levin, & Steven E. Plotkin, *Plug-in Electric Vehicle Policy Effectiveness: Literature Review*, Argonne National Laboratory: Energy Systems Division (May 2016), https://perma.cc/2CNP-AB9E (last visited June 19, 2018).

ENDNOTES

1   BENJAMIN JACOBS, ZERO EMISSION VEHICLE MUNICIPAL HANDBOOK: A LAND USE GUIDE FOR CITIES AND TOWNS 12-13 (Rhode Island Office of Statewide Planning May 2017), http://perma.cc/8URS-3V6M (last visited May 31, 2018).
2   *Id.* at 15-16.
3   *Id.*
4   *Id.* at 14.
5   *Id.*
6   *Id.*
7   *Id.* at 21-22.
8   Office of Transportation and Air Quality, *Fast Facts: U.S. Transportation Sector Greenhouse Gas Emission 1990-2015*, EPA (July 2017), https://perma.cc/PTC8-GMWZ (last visited June 19, 2018); Office of Transportation and Air Quality, *Greenhouse Gas Emissions from Typical Passenger Vehicle*, EPA (March 2018), https://perma.cc/PL5P-A5UT (last visited June 19, 2018).
9   Office of Transportation and Air Quality, *supra* note 8; Alternative Fuels Data Center, *Benefits and Considerations of Electricity as a Vehicle Fuel*, DOE, https://perma.cc/24T4-LA4Q (last visited June 4, 2018).
10 Luke Tonachel, *Study: Electric Vehicles Can Dramatically Reduce Carbon Pollution from Transportion, and Improve Air Quality*, NRDC (Sept. 17, 2015), http://perma.cc/E3ZJ-PATE.
11 David Block et al., *Electric Vehicle Sales for 2014 and Future Projections*, Electric Vehicle Transportation Center, at 8 (March 30, 2015), https://perma.cc/FND8-X9V8.
12 Yan Zhou et al., *Plug-in Electric Vehicle Policy Effectiveness: Literature Review*, Argonne National Laboratory: Energy Systems Division, at 20 (May 2016), https://perma.cc/2CNP-AB9E (last visited June 19, 2018).
13 *Id.*
14 City of Chelan, WA, Chelan Municipal Code § 17.63.020 (2011).
15 *Id.*
16 *Id.*; JACOBS, *supra* note 1, at 15.
17 City of Atlanta, GA, Ordinance No. 2014-53 (14-01278) (May 12, 2014).
18 City of Atlanta, GA, Code of Ordinances § 16-29.0019(56) (2014).

PROPERTY ASSESSED CLEAN ENERGY PROGRAM

Kyler Massner (author)
Jonathan Rosenbloom & Christopher Duerksen (editors)

INTRODUCTION

Property Assessed Clean Energy (PACE) programs provide a mechanism for owners of private property to finance low-cost, long-term funding for renewable energy (RE) and energy efficiency (EE) improvements. Most importantly, PACE programs offer homeowners an opportunity to take advantage of renewable energies without having a substantial upfront cost.[1] PACE programs are structured to provide 100% financing of a RE or EE project's cost. That upfront cost is secured by the property, backed by the local government, and repaid by the property owner through an additional assessment on the owner's taxes for a term of up to 20 years.[2] Local governments can make PACE available to both commercial (C-PACE) and residential properties (R-PACE). As of 2017, over 150,000 homeowners have made approximately $4 billion in RE and EE improvements through PACE programs.[3]

In most jurisdictions, state governments must enact enabling legislation authorizing local governments to offer PACE financing and form PACE assessment districts that recognize RE and EE developments as public "goods."[4] After the districts are created, local governments establish special assessments for utilities that are financed through property assessments and similar collection procedures.[5] These assessment districts benefit local governments by protecting or insulating the government's debt rating from being impacted by the PACE financing program.[6] Participation in the district is typically voluntary, allowing property owners to opt-in.[7]

PACE financing "debt" (evidencing the upfront costs) attaches to the property as a lien.[8] Securing the debt by a lien on the property has three distinct advantages. First, it reduces the initial financial burden by spreading out repayment over many years.[9] Second, the repayment obligation can transfer with the property, and thus is not required to be resolved prior to a subsequent sale.[10] Third, it provides owners assurance that their investment is protected, thus reducing apprehension of investing in RE and EE improvements.[11]

Sustainable Development Code: Climate Change

Commercial and residential electricity usage accounts for roughly 32% of all electricity consumed in the U.S., with much of the generation coming from fossil fuel dependent and pollution emitting assets.[12] PACE financing reduces reliance on fossil fuels by encouraging the adoption of RE and EE technology. By utilizing PACE to install RE and EE technology, property owners can reduce utility bills and their carbon footprint.[13] Communities that take advantage of PACE can encourage RE and EE developments without obligating money from the local governments' general fund.[14]

In addition to environmental benefits, PACE financing programs have a positive economic impact in the local community. For example, PACE programs financed roughly $4 billion in clean energy projects and have created approximately 35,000 new jobs since 2008.[15] For every $1 million spent, 60 jobs are created and $10 million in gross economic output and $1 million in combined tax revenue is generated.[16] PACE programs, as a result, can help stimulate and stabilize the local economy, while reducing energy costs and levels of greenhouse gases.[17]

Since PACE's inception, concerns over the PACE financing program have been voiced by bank regulators and market entities concerning the security of collateral in the event of a default. As recently as December 7, 2017 the U.S. Dep't of Housing and Urban Development stated that properties encumbered by PACE liens are no longer eligible for FHA-insured forward mortgages.[18] Analysis shows, however, that concerns of increased risk may be counterbalanced by the benefits which the PACE program provides.[19] For example, PACE programs can stimulate the local economy, generate tax revenue, and install improvements on a home which increases its value and marketability, thus mitigating some concerns of the banking industry.[20]

EXAMPLES

Thirty-three states as well as D.C. offer some form of PACE financing.[21] Below are several examples of both residential and commercial PACE programs established by local governments in these 33 states.

Los Angeles County, CA

In 2010, Los Angeles County began offering Property Assessed Clean Energy (PACE) programs to both Commercial (C-PACE) and Residential (R-PACE) owners.[22] The program, Los Angeles County Energy Program (LACEP), has been increasingly expanded and is intended to encourage installation of RE and EE improvements without the need for a large down payment.[23] Improvements can enhance property values, lower energy bills, reduce greenhouse gas emissions, and spur development of "green" economy jobs within the County.[24] All improvements that can be permanently fixed to the property and are proven to save energy or produce renewable energy are eligible for approval.[25] Such projects include high efficiency cooling and heating systems, electric vehicle charging stations, insulation, cool roofs, solar panels, smart irrigation systems, and more.[26]

Under the County's PACE program, property owners contract with one of two County approved program administrators to finance the improvements.[27] The County is then authorized to issue bonds to these administrators under strict underwriting requirements.[28] The interest rates on these bonds are determined by conditions in the taxable bond market at the time of sale.[29] Once approved, the bonds are used to secure financing for RE and EE projects, which then are repaid to the municipality via an assessment on the owner's property taxes.[30]

As of 2018, 87 of the 88 cities in Los Angeles County have adopted PACE through participation in LACEP.[31] Participation is voluntary, and municipalities can join by adopting a resolution that authorizes property owners to apply for PACE financing.[32] Santa Clarita's R-PACE program became active in 2015 and as of November of 2016, participation resulted in 472 completed residential PACE projects, 55 PACE registered contractors, a total value of $12 million spent, with an estimated 250 future PACE projects.[33]

To view the program see Los Angeles County Energy Program (LACEP) Program Report (Jan. 24, 2017).

San Francisco, CA

San Francisco adopted and enabled PACE in 2010. Its program authorized the collection and levying of Special Taxes within the San Francisco Special Tax District. In 2011, San Francisco authorized the issuance and sale of special tax bonds for the PACE program to provide financing and refinancing for the acquisition and installation of RE and EE improvements on public and private property.[34] Under San Francisco's C-PACE program, known as "GreenFinanceSF", commercial property owners can secure 100% financing for RE and EE improvements.[35] GreenFinanceSF seeks to overcome the biggest hurdle to commercial energy upgrades by providing commercial property owners an avenue of finance to make improvements, while saving energy and money, and increasing both the property's value and attractiveness to new or current tenants.[36] Once the RE and EE improvements have been made, property owners are protected from utility cost increases and can share the benefits with their tenants.[37]

San Francisco's PACE program uses the "open market" PACE model which allows prospective developers to finance a project with a qualified capital provider of their choice.[38] An "open market" system permits an individual owner to target lenders and negotiate financing terms. The City then collects repayments from a special tax lien on the property and then pays the lender like any traditional PACE program.[39] To be eligible, property owners must be current on all property taxes, assessments, and liens for the preceding three years.[40] Furthermore, a professional energy and/or water audit must be conducted (for ordinances requiring and incentivizing commercial energy audits see Energy Benchmarking, Auditing, and Upgrading brief).[41] If a property owner receives financing for a RE system, the owner is also required to implement EE measures that result in at least a 10% improvement in energy performance.[42]

An example of GreenFinanceSF at work is Pier 1 in northeast San Francisco.[43] Under GreenFinanceSF, Prologis Inc. chose Clean Fund as their PACE capital provider, securing $1,400,000 in funds.[44] This made possible the installation of 200 kW of solar electric panels, retro-commissioned utilities, and an extensive lighting upgrade on Pier 1.[45] These energy savings allowed the pier to cut energy costs by 32% with an approximate $98,000 of savings per year and an estimated creation of nearly 30 jobs and $3.7 million in economic development as calculated by the U.S. Department of Commerce.[46]

To view the provisions see San Francisco, CA, Ordinance No. 118-16 (July 31, 2016); San Francisco, Cal., Ordinance No. 308-11 (July 26, 2011).

ADDITIONAL EXAMPLES

Miami-Dade County, Fla., Code of Ordinances §§ 2-2079-2091 (2018) (creating a county-wide PACE program that includes unincorporated municipalities).

See also Miami-Dade Green, *Miami-Dade County Energy Efficiency and Renewable Energy Finance Program: Update of Existing Programs*, Office of Sustainability (November 2013), https://perma.cc/KF4K-KY3M (last visited May 29, 2018) (recommendations and updates regarding PACE programs).

St. Louis, Mo. Code of Ordinances § 3.120.010-.060 (2011); *Set the PACE St. Louis*, St. Louis, MO, https://perma.cc/7WKG-UBT2 (last visited May 29, 2018) (using Ygrene LLC, to administer the St. Louis program).

District of Columbia - Audi Field

The District of Columbia began its PACE program in 2010 ("DCPACE").[47] Pursuant to DCPACE, the initial investment is provided by private lenders, and the loan is repaid through an assessment on property taxes. D.C.'s Department of Energy and Environment administers the DCPACE program.[48] To date, 16 projects have received approximately $34,000,000.

Opening in 2018, the D.C. United Audi Field, is the latest product of DCPACE.[49] The owners of the project, D.C. United, secured a $25,000,000 investment from a locally based private lender to deploy an 884 kW solar PV array, a storm water retention system at the stadium and other energy efficiency measures.[50] The solar array offsets the stadium's energy consumption by a third, producing approximately 1,000,000 kWh of solar electricity and reducing carbon emissions by 820 tons annually.

To view the provisions see Energy Efficiency Financing Amendment Act of 2012, 17R D.C. Code §§ 8-1778.01-.02. .21-.31, .41-.48 (2012); 8 D.C. Code §§ 47-895.31-.35 (2015).

ADDITIONAL RESOURCES

Pass PACE Legislation in my State, PACENation, https://perma.cc/WBJ9-FASQ (last visited June 19, 2018) (includes tools, tips and legal framework for enacting PACE).

NC Clean Energy Technology Center, *Programs,* DOE, https://perma.cc/P7N3-GVU8 (last visited June 19, 2018) (database of state incentives for renewables & efficiency upgrades).

Fact Sheet Series on Financing Renewable Energy Projects, NREL, https://perma.cc/7HB6-25XB (last visited May 22, 2018) (national renewable energy laboratory PACE factsheet).

Best Practice Guidelines for Residential PACE Financing Programs, DOE, (November 18, 2016), https://perma.cc/Y5XP-5PQV (last visited May 18, 2018).

Real Estate Review Journal, Charlene Vanlier Heydinger, 44 No. 1 Real Estate Review Journal ART 3 (identifying how PACE financing overcomes barriers to RE and EE deployment).

ENDNOTES

1   Office of Energy Efficiency & Renewable Energy, *Property Assessed Clean Energy Programs*, DOE, https://perma.cc/RRE3-YYCS (in this brief RE and EE are given a broad meanings ranging from traditional home improvements, *e.g.*, new windows or insulation, to solar and wind projects) (last visited June 14, 2018).

2   *Id.*; *see* Lindsay Breslau, Michael Croweak, Alan Witt, *Batteries Included: Incentivizing Energy Storage*, 17 Sustainable Dev. L. & Pol'y 29, 36 (2017) (suggesting PACE financing to increase deployment of energy storage projects).

3   *Property Assessed Clean Energy Programs, supra* note 1.

4   *Best Practice Guidelines for Residential PACE Financing Programs*, DOE 1 (Nov. 18, 2016), https://perma.cc/Y5XP-5PQV.

5   *Id.*at 2-3; *see* National Renewable Energy Laboratory, *Energy Analysis: Property Assessed Clean Energy (PACE) Financing of Renewables and* Efficiency, DOE tbl.1, (July 2010), https://perma.cc/7HB6-25XB (last visited on May 18, 2018) (*Local governments have a variety of options to finance special assessments*).

6   National Renewable Energy Laboratory, *supra* note 5, at 1-2.

7   *Id.*

8   *Id.*

9   *Id.*

10   *Id.*

11   *Id.*

12   *Sources of Greenhouse Gas Emissions*, EPA, https://perma.cc/DEK2-GY3R (last visited Mar 17, 2018).

13   *Best Practice Guidelines for Residential PACE Financing Programs, supra* note 4, at 4-5.

14   *Energy Analysis: Property Assessed Clean Energy (PACE) Financing of Renewables and Efficiency, supra* note 5, at 2.

15   Katie Fehrenbacher, *As Pace Financing Grows Up, the Industry Grapples with Lending Standards and Consumer Protections*, GreenTech Media, (Mar. 29, 2017), https://perma.cc/M3CL-44AB (detailing the benefits of PACE programs nationwide; curbing enthusiasm by detailing some predatory lending practices).

16   ECONorthwest, *Economic Impact Analysis of Property Assessed Clean Energy Programs (PACE)* 1-2, PACENow, (April 2011), https://perma.cc/T8VP-FG5Z (last visited May 18, 2018).

17   *Id.*

18   Letter from Dana T. Wade, General Deputy Assistant Secretary for Housing, to FHA-approved Mortgagees et. al., Property Assessed Clean Energy (PACE), at 2-3 (Dec. 7, 2017), https://perma.cc/N5NH-CBMK.

19   ECONorthwest, *supra* note 16 at 13-15.

20   *Id.*

21   *See* Christopher J. Schreiber & David I. Cisar, *Emerging Issues: Residential Pace Loans and Bankruptcy*, Am. Bankr. Inst. J., Feb. 2018, at 32; *see also* NC Clean Energy Technology Center, *Programs*, DOE, https://perma.cc/P7N3-GVU8 (last visited June 19, 2018) (indexing current PACE programs throughout the United States).

22   *Los Angeles County Energy Program (LACEP): Program Report*, Cty. of Los Angeles 14 (Jan. 24, 2017), https://perma.cc/88M2-7PV6.

23   *Id.*at 2.

24   *Id.* at 15.

25   *Id.*

26   *See id.* app. B.

27   *Id.* at 16.

28   *Id.* at 19-22.

29   *Id.* at 20.

30   *Id.* at 14.

31   *Los Angeles County Pace*, Cty. of Los Angeles, https://perma.cc/P2GL-L5R2 (last visited May 22, 2018).

32   Cty. of Los Angeles, *supra* note 22, at 14.

33   City of Santa Clarita Agenda Report, Cmty. Dev., 1-2 (Cal. 2017), https://perma.cc/2CPD-VBDK.

34   San Francisco, CA, Ordinance No. 016-10 (Feb. 8, 2010); San Francisco, CA, Ordinance No. 308-11 (July 26, 2011).

35   SF Env't, *GreenFinanceSF: Commercial PACE program*, San Francisco Dep't of the Env't, https://perma. cc/Z24X-CE87 (last visited May 22, 2018).

36   SF Env't, *PACE Financing for Prologis at Pier 1*, San Francisco Dep't of the Env't, https://perma.cc/ SF3Q-FDYB (last visited May 22, 2018).

37   *See id.*

38   *City of San Francisco - GreenFinanceSF*, DOE, https://perma.cc/6W4F-EUPU (last visited May 22, 2018).

39   *Id.*

40   *Id.*

41   *Id.*

42   *Id.*

43   SF Env't, *PACE Financing for Prologis at Pier 1*, San Francisco Dep't of the Env't, https://perma. cc/5W2P-JPN9 (last visited May 22, 2018).

44   *Id.*

45   *Id.*

46   *Id.*

47   D.C. Code Ann. § 8-1778.41 (2012).

48   *Property Assessed Clean Energy*, DOE, https://perma.cc/U9UH-Y7RK (last visited June 14, 2018).

49   DCPACE, *D.C. United Audi Field*, Urban Ingenuity, https://perma.cc/8599-RFY9 (last visited June 14, 2018).

50   *Id.*

PROPERTY TAX EXEMPTIONS FOR RENEWABLE ENERGY SYSTEMS

Brandon Hanson (author)
Jonathan Rosenbloom & Christopher Duerksen (editors)

INTRODUCTION

Renewable energy generation systems, like solar panels and geothermal, have the potential to increase property values.[1] However, even with incentives and rebates, some property owners may be hesitant to install renewable energy generation systems because the increased value often translates into higher property taxes. An ordinance that exempts the value augmented to property by the addition of renewable energy generation systems from property taxes makes owning solar, wind, geothermal or other renewable energy sources less burdensome and may encourage increased renewable systems.

This ordinance entices people to install and own renewable energy systems by exempting the added value of renewable energy systems from property tax assessments. Local governments can draft this ordinance to give homeowners the ability to have the additional value of property created by a renewable energy generation system, exempted from the taxable value of their property.[2] This would make the taxable value of the home the same as before the renewable energy system was installed.[3] This ordinance may contain the types of renewable energy systems that qualify for the property tax exemption (typically a very inclusive list), the maximum amount that is exempt, and the proof necessary to show the amount exempted (usually determined by an assessor or percentage of the renewable energy system).[4] Ordinances can be created to include exemptions for residential and/or commercial property, making them beneficial to business owners as well residential property owners.[5]

Over half the states have some form of tax exemption, either on the state level or a local opt-in or out program.[6] Some states mandate that local governments provide some form of tax exemption for solar panels that can be adjusted by the local government.[7] A full list of tax incentives for property and other taxes can be found in the *Green Tax Incentive Compendium*.[8]

Providing residential and/or commercial property tax exemptions for renewable energy systems can promote local economies, encourage healthy living, and reduce greenhouse gas (GHG) emissions and other pollutants involved with the burning of fossil fuels.[9] Renewable energy sources help minimize dependence on fossil fuels, that create air pollution and emit GHGs.[10] Coal, natural gas, and petroleum make up 99% of carbon dioxide emissions from electrical energy production.[11] These GHG emissions have created major health concerns, with new residential and commercial property owners utilizing renewable energy generation systems reliance on fossil fuels can drop, benefiting local communities.[12] Burning fewer fossil fuels, reduces GHG's and other air pollutant emissions, thereby helping to mitigate climate changing impacts and improve air quality.[13]

Local governments that provide tax exemptions for renewable energy systems make it more affordable to own and keep renewable energy systems and homes containing those systems. Higher property taxes make certain areas less desirable to live in and more expensive to open a business. Local governments that provide exemptions for the assessed added value of property after the installation of a renewable energy system create a more desirable location for perspective home and business owners.[14] With more municipalities offering tax exemptions for the value added by renewable energy systems, consumers may be inclined to move to these jurisdictions, while the lack of an exemption may lead individuals to avoid these locations.[15] Increasing property value has become a negative for many homeowners, as an increase in value means an increase in taxes. Failure to maintain property can lead to a decrease in value of the neighborhood that may lower property values in the area.[16] Having an ordinance that exempts the additional value encourages property owners to improve property, which can increase the value of an area as whole.

With more people looking to install and maintain renewable energy sources, new jobs for the installation and maintenance will also be created.[17] The labor-intensive fields of renewable energy have been growing and will continue to grow with more renewable energy generation systems being built.[18] The added jobs can have a ripple effect on local markets by bringing in more people who in turn spend money at local stores and restaurants.[19] Furthermore, the addition of more renewable energy sources helps stabilize utility prices. By using an inexhaustible resource, such as wind and solar, the prices of electricity are less volatile to the shifting prices of fossil fuels, giving consumers a more consistent energy bill.[20] The use of renewable energy gives the community more predictable energy bills and can drastically lower energy costs for homeowners.[21]

EXAMPLES

Pearisburg, VA

Pearisburg exempts certified solar energy equipment, facilities, and devices from property taxation.[22] To receive the exemption for real or personal property, the local authority (in this ordinance the county Building Department) must designate the solar equipment as certified pursuant to the specific criteria made by the Virginia (State) Board of Housing and Community development.[23] The equipment must be used to collect or create energy that will replace energy consumption from fossil fuels.[24] The City's exemption applies to any person residing in Pearisburg, and the equipment can be fully or partially exempt from taxation.[25] The amount of the exemption is determined by deducting the property tax amount of the solar equipment from the real property tax due on the property to which the equipment is attached.[26] The Ordinance also presumes that the value of the qualifying solar equipment is not less than the cost of purchasing and installing the equipment.[27] The tax exemption is effective when the property is first assessed following the installation of the system.[28]

To view this provision see Pearisburg, VA, Code of Ordinances § 66-37 (2014).

Winder, GA

Winder exempts from property tax assessment all tangible

ADDITIONAL EXAMPLES

Kenai Peninsula Borough, AK, Code of Ordinances tit. 5 § 5.12.101 (2018) (exempting residential renewable energy systems from real property tax.

Vernon, CT, Code of Ordinances § 12-5 (2011) (providing a tax exemption for the value of a renewable energy system attached to property that exceeds the amount of such property with conventional energy systems).

Cheshire, CT, Code of Ordinances §§ 17-2 - 17-5 (2017) (providing property tax exemptions for solar heating or cooling systems; those wanting to qualify must apply and the exemption is good for 15 years after construction).

Warwick, RI, Code of Ordinances § 74-52 (2017) (creating a tax exemption for additional value added by a renewable energy system).

property used for solar energy heating or cooling, consisting of equipment used to manufacture of solar energy systems.[29] This includes controls, tanks, pumps and other equipment used for utilizing solar energy, but does not include walls or other structures modified to capture energy from the sun.[30] Meaning things that would ordinarily be part of a building, such as walls that are painted to harness or repel solar energy (heat), are not considered for tax exemptions. This ordinance adds some limitations to the amount of tax exemptions, but still applies to a broad range of improvements, including water heating and drying devices.[31]

To view this provision see Winder, GA, Code of Ordinances § 29-10 (2017).

ENDNOTES

1   Samuel Dastrup et. al., *Understanding the Solar Home Price Premium: Electricity Generation and Green Social Status* (Nat'l Bureau of Econ. Research July, 2011) (finding that the addition of solar panels in San Diego raises property value over $22,000) https://perma.cc/8GB5-R4SX.

2   Jacob Marsh, *Solar Property Tax Exemptions: Are They Available Where You Live?*, Energy Sage (Aug. 22 2017), https://perma.cc/H428-ZTYP (last visited May 29, 2018).

3   *See id.*

4   Warwick, RI, Code of Ordinances § 74-52 (Dec. 20, 2017).

5   Marsh, *supra* note 2.

6   Irina Rodina & Shaun A. Goho, *The Solar Property Tax Exemption in Massachusetts: Interpretation of Existing Law and Recommendations for Amendments*, Emmett Environmental Law & Policy Clinic, Harvard Law School (July 2013).

7   *Id.*

8   Jerome L. Garciano, Robinson & Cole L.L.P., *Green Tax Incentive Compendium* (Jan. 1, 2018), https://perma.cc/NHY4-K9F3.

9   UCS, *Benefits of Renewable Energy Use* (Dec. 20, 2017), https://perma.cc/XY55-XGYP (last visited June 1, 2018).

10   EPA, *State Renewable Energy Resources*, https://perma.cc/Z35K-C5JV (last visited May 25, 2018).

11   U.S. Energy Information Admin., *How much of U.S. carbon dioxide emissions are associated with electricity generation?*, (May 10, 2017) https://perma.cc/P6X4-VQ2V.

12   *Id.*

13   Marsh, *supra* note 2.

14   FindLaw, *Impact of Changing Property Values on Property Taxes*, https://perma.cc/XC34-9WG5 (last visited June 4, 2018).

15   *Id.*

16   *Id.*

17   *Id.*

18   Austin Brown et al., *Estimating Renewable Energy Economic Potential in the United States: Methodology and Initial Results* (NREL 2016) https://perma.cc/FY2Q-SNUQ.

19   *Id.*

20   UCS, *supra* note 9

21   Hannah West, *Long and Short Term Benefits of Solar Power at Home*, Proud Green Home (July 9, 2015), https://perma.cc/SH9T-JZES (last visited June 1, 2018) (noting a savings of $60,000 in 25 years).

22   Pearisburg, VA, Code of Ordinances § 66-37 (a) (2014).

23   *Id.* at (b).

24   *Id.*

25   *Id.* at (c).
26   *Id.*
27   *Id.* at (d).
28   *Id.* at (c).
29   Winder, GA, Code of Ordinances § 29-10 (2017).
30   *Id.* at (b).
31   *Id.* at (c).

RECYCLE, SALVAGE AND REUSE BUILDING MATERIALS

Brandon Hanson (author)

Jonathan Rosenbloom & Christopher Duerksen (editors)

INTRODUCTION

Every year the construction industry in the U.S. produces over 160 million tons of construction and demolition materials.[1] Most of these materials are sent to landfills, and each year the amount of space needed for landfills grows. The landfill space needed for this material could be drastically reduced by salvaging and recycling.[2] Many materials, such as brick, wood, concrete, roofing materials, asphalt, and metals, can have reuse purposes and can be recycled to generate new raw materials.[3] Local governments should enact ordinances that require or encourage a specific minimum percentage of materials removed from buildings during demolition to be diverted from a landfill by either reuse, recycling, or other ways.

Ordinances addressing construction or demolition should be drafted to require a minimum percentage of total waste or an amount per square foot of building space being erected or razed be diverted from landfills. Meeting diversion rates encourages developers to use more sustainable practices and to use material that can easily be reused or salvaged before the demolition of a structure. Local governments can impose monetary sanctions for not complying with the ordinances and/or suspend or revoke a developers' permit to build. Another way to reach a similar goal, is to require specific materials in construction and demolition to be recycled or salvaged. The specific materials can be added to a diversion rate ordinance or be enacted separately.

Local governments may also set specific rates or percentages for waste to be salvaged rather than recycled. Alternatively, local governments could create incentives for de-construction and salvaged materials, as opposed to demolition. Deconstruction is the process of taking apart existing constructions rather than leveling and separating materials. This makes it easier to salvage, reuse, and recycle materials, leading to greater waste diversion rates from landfills.[4]

EFFECTS

The amount of debris estimated to be generated in 50 years by the construction and demolition industries is over 3 billion tons.[5] Encouraging or requiring salvaging and recycling of that material can have a large impact on the environment and communities. Enacting an ordinance that addresses building materials and demolition helps keep materials out of landfills.[6] Also, reusing building materials saves energy, compared to creating new building materials.[7] The amount of energy needed to produce building material from raw materials is drastically higher than reusing or recycling materials from structures being razed.[8] With lower energy consumption, greenhouse gas (GHG) emissions drop, helping to mitigate climate change and improve air quality.[9] By encouraging or requiring reuse, salvaging, and recycling, local governments can reduce the building and demolition industry's impact on the environment.

This ordinance may also have a positive economic impact. Local governments are often touting the number of jobs various projects create. Because deconstruction has been shown to require more labor and time than demolition, it requires more jobs for an equivalent project. Another area that could see an increase in employment is the recycling and reuse industry. If local governments require a certain percentages to be diverted from landfills, more will be sent to recycling and reuse centers.[10] This will require more employees to sort and process the materials.[11]

EXAMPLES

San Francisco, CA

The City of San Francisco requires a 65% minimum diversion rate of construction and demolition debris from landfills.[12] Any person, firm, company wanting to demolish a structure is required to submit a plan that lists the materials expected to be a part of the demolition and 65% of the materials are restricted from going to a landfill.[13] Failure to comply with the ordinance can result in suspension of licenses and permits required for operation of facilities and machinery related to the demolition and destruction industry.[14] The City can also impose monetary penalties of up to $1,000 for each day of non-conformity.[15] Enforcement of the code is implemented by the City's director of the environmental code and can inspect any property registered for demolition to insure compliance of the diversion rate.[16] The City is also planning to increase the percentage of diversion over the upcoming years in an effort to reach zero waste.

To view this provision see San Francisco, CA, Environmental Code §§ 1400-17 (2006).

Austin, TX

For certain buildings (described in § 25-11-39), Austin set: 1) a maximum weight of waste that may be disposed of per square foot, and 2) a minimum of 50% total waste diversion rate from landfills. Austin will decrease the maximum weight of waste per square foot 2.5 pounds to .5 pounds by 2030.[17] Subject to subsequent Council approval, the City also codified an increase in percentage of waste to be diverted from landfills: 50% will increase to 75% by 2020 and 95% by 2030. This is the minimum diversion rate of materials generated by the construction and demolition.[18] Failure to comply with the code is a class C Misdemeanor and can contain monetary fines. The diversion rate is determined by the local Resource Recovery Department and materials can be given to a qualified processor to meet the requirements.[19] The processor weighs and determines if the material diversion rate is met.[20] Processors must file reports to the department to ensure compliance with the current diversion rate. Waivers may be requested and given with the approval of the director of the Resource Recovery Department.[21]

To view this provision see Austin, TX, Code of Ordinances §§ 15.6.150 – 6.170 (2016).

ADDITIONAL EXAMPLES

Portland, OR, City Code § 17.106 (2016) (setting deconstruction requirements for specific building types, making the reuse of materials easier).

Orange County, NC, Code of Ordinances § 34.73 (2002) (mandating specific types of construction materials to be recycled, demolition requires documentation of recycling).

Northbrook, IL, Code of Ordinances §§ 6.241 – 6.251 (2008) (setting a 75% diversion rate for construction and demolition materials from landfills).

Arroyo Grande, CA, Code of Ordinances § 8.32.200 (2017) (requiring deconstruction or salvage to fullest extent possible before following mandatory recycling plan).

Minneapolis, MN, Code of Ordinances § 527.260 (2018) (sustainable building practices including deconstruction plans are a primary consideration for demolition approval).

ENDNOTES

1   EPA, OSWER Innovation Project Success Story: DECONSTRUCTION (Nov. 2009), https://perma.cc/E97K-LQA5.
2   EPA, Recover Your Resources: Reduce, Reuse, and Recycle Construction and Demolition Materials at Land Revitalization Projects (Oct. 2009), https://perma.cc/FTF5-K7R5.
3   EPA, *supra* note 2.
4   *Id.*
5   A.C. Nelson, *Toward a New Metropolis: The Opportunity to Rebuild America* (Dec. 2004) (discussion paper prepared for The Brookings Institution Metropolitan Policy Program).
6   EPA, *supra* note 2.
7   Stephanie Boyd et al., *Deconstructing Deconstruction: Is a Ton of Material Worth a Ton of Work?*, 5 Sustainability J. of Record 391 (Dec. 2012).
8   *Id.*
9   *Id.*
10 Construction & Demolition Recycling Association, *This is Where the Real Work Begins*, https://perma.cc/24WA-BLSU (last visited June 7, 2018) (noting the C&D recycling industry will create over 28,000 jobs).
11 *Id.*
12 San Francisco, CA, Environmental Code §1402 (b) (2006).
13 *Id.*
14 *Id.* § 1410 (c).
15 *Id.* § 1410 (e).
16 *Id.* § 1410 (a), (b).
17 Austin, TX, Code of Ordinances §§ 15.6.151- 6.152 (2016).
18 *Id.*
19 *Id.* § 15.6.155.
20 *Id.*
21 *Id.* § 15-6-1 (12); *id.* § 15.6.156.

RECYCLING IN MULTI-FAMILY AND COMMERCIAL BUILDINGS

Tyler Adams (author)
Jonathan Rosenbloom & Christopher Duerksen (editors)

INTRODUCTION

Municipalities have increasingly turned to recycling initiatives as a way to further sustainability goals. An area ripe for opportunity is multi-family and commercial buildings which are often left out of such initiatives. Because these building produce large amounts of waste and often do not have or require recycling, large amounts of recyclable material continue to go to landfills. Commercial and multi-family waste production is nearly twice that of residential, thus implementing recycling programs in these buildings would positively impact a municipality's diversion rate (the measure of the amount of recyclable material avoiding landfills).[1]

Due to the temporary nature of their occupants, the recycling challenges commercial and multi-family buildings face are often different from single-family dwellings.[2] Local governments have begun requiring commercial and multi-family property owners to offer tenant/occupants the opportunity to recycle. Previously, commercial and multi-family buildings were only required to provide the standard waste disposal bins.[3] Pursuant to some ordinances, they must provide separate bins that divide recyclable and non-recyclable materials. Often times the buildings are also required to provide information to tenant/occupants as well a report to the appropriate municipal department regarding recycling efforts. In addition, some jurisdictions phase buildings into compliance based on size, use and/or amount of waste generated.[4]

EFFECTS

Ensuring recycling in multi-family and commercial buildings has the potential to confer numerous benefits. Trash can end up in a variety of places once it is thrown away, but one of the most common places is landfills. The decomposition of organic materials in landfills produces landfill gas (LFG) which is composed of about 50% methane and 50% carbon dioxide, both of which are potent greenhouse gases (GHGs).[5] Landfills are the third largest source of human-related methane emissions. By reducing the amount of waste sent to landfills through diversion, the emission of GHGs is decreased.[6] Some waste also makes its way to incinerators where it is burned and turned to ash.[7] Although incinerators can generate electricity, the burning process produces gases like nitrogen oxides and sulfur dioxide both of which cause smog.[8] The emissions of these gases can also be reduced by increasing recycling in multi-family and commercial buildings. Further, recycled material is able to be reused to produce similar products which decreases the consumption of natural resources.[9]

In addition to its environmental benefits, increased recycling can benefit a municipality economically. According to the 2016 Recycling Information Report, recycling generated over six and a half billion dollars in local and state tax revenue in 2007.[10] Further, "[s]tudies have shown that for every one job in waste management there are four jobs in recycling. After the recycling process, even more jobs are created for making new goods out of the recycled materials."[11] Owners of commercial or multi-family buildings are also able to benefit due to recycling collection services often being less expensive than regular waste collection services.[12]

EXAMPLES

Citrus Heights, CA

This ordinance states that any business or multi-family residential property owner that generates four or more cubic yards of commercial solid waste per week or owns a multi-family property with five or more dwelling units is a "covered generator" and is required to obtain and maintain recycling collection services.[13] "Covered generators" must also obtain organics recycling services with some businesses being required to separate food scraps from green materials.[14] "Covered generators" can either enter into service agreements with authorized haulers or do self-hauling in accordance with the ordinance.[15] In addition, recyclable material containers are to be provided in multi-family residential rental units as well as maintenance and work areas where recyclable materials may be collected or stored.[16] A sign stating where an employee or tenant can find these containers and what is to be separated must also be posted.[17] Tenants in a multi-family building are responsible for separating their recyclable materials and a building owner cannot be cited for their noncompliance.[18]

To view the provision see Citrus Heights, CA, Code of Ordinances § 74-128 (2016).

Austin, TX

As of October 1, 2016, all owners or managers of multi-family residential properties with at least five units must ensure access to on-site recycling services and be in compliance with Austin's Universal Recycling program.[19] All premises for which all or part is used for non-residential use must be in compliance by October 1, 2017.[20] At a minimum, the materials to be separated must include paper, plastic, aluminum, corrugated cardboard, and glass.[21] Organic material must also be collected if the premises contains a food enterprise.[22] A manager or owner of a property can remove the recyclable material by either contracting with a City-licensed hauler or self-hauling. When self-hauling, they have the option of not only taking the materials to recovery facilities, but also to urban and rural farms or ranches and food banks.[23] In addition to a recycling plan that needs to be submitted to the City, responsible parties are also required to provide recycling information and instructions to all tenants and employees annually.[24]

To view the provision see Austin, TX, Code of Ordinances § 15-6-91 (2014).

ENDNOTES

1   *Multifamily and Commercial Recycling*, City of Orlando, http://perma.cc/YH2B-3WC4 (last visited May 29, 2018).

2   Recycle Ann Arbor, Multi-Family Recycling Incentive Pilot Program 3 (Sep. 8, 2017), http://perma.cc/E5FL-CQKX.

3   City of Orlando, *supra* note 1.

4   *Id.*

5   *Landfill Methane Outreach Program (LMOP)*, EPA, http://perma.cc/E44Q-J4XW (last visited May, 29 2018).

6   *Id.*

7   Dan Kulpinski, *Human Footprint: Where Does All the Stuff Go?*, Nat'l Geographic, http://perma.cc/8PME-DQRP (last visited May, 25 2018).

8   *Id.*

9   *Id.*

10   *The Economic Benefits of Recycling*, Ever Green Environmental (Nov. 9, 2017), http://perma.cc/UT3H-MHW3.

11   *The Benefits of Recycling*, Renewable Resources Coalition (Dec. 15, 2016), http://perma.cc/HDR6-MGUU.

12   City of Orlando, *supra* note 1.

13   Citrus Heights, CA, Code of Ordinances § 74-128 (2016).

14   *Id.*

15   *Id.* § 74-129.

16   *Id.*

17   *Id.*

18   *Id.* § 74-137.

19   Austin, TX, Code of Ordinances § 15-6-91 (2014).

20   *Id.*

21   *Id.* § 15-6-92.

22   *Id.*

23   *Id.*

24   *Id.* §§ 15-6-93, 15-6-101.

RENEWABLE ENERGY WITH INCENTIVES

Brandon Hanson (author)
Jonathan Rosenbloom & Christopher Duerksen (editors)

INTRODUCTION

Coal, natural gas, and petroleum make up 99% of carbon dioxide emissions from electrical energy production.[1] Renewable energy sources help minimize dependence on fossil fuels that create air pollution and emit greenhouse gases (GHG).[2] Local governments can provide incentives for residential and commercial property owners to move away from traditional energy sources and toward renewable sources. Doing so is a prodigious way to help reduce emissions from fossil fuels. Common renewable energy generation systems are wind, solar, hydroelectric and geothermal. This ordinance should be drafted in a way to leverage the most beneficial renewable sources based on the local government's location. For example, local governments in the sun-drenched portions of the southwest may seek to take advantage of and create incentives for solar energy.[3] Solar panels can be added to existing structures easily and with little effect to the building, making solar an ideal source of renewable energy for those with an abundance of sun.

In drafting this ordinance, local governments have a variety of options for creating incentives, including offering rebates on purchasing equipment, tax incentives, height allowances, setback and area-based incentives, expedited permitting, and others.[4] One incentive would allow net metering. Net metering measures the amount of energy produced by a renewable energy generation system. If more energy is produced than needed, a credit can be issued to the resident. The credits can be utilized to pay for utility bills in months when less energy is produced (see Zero Net Energy Buildings).[5] Another important incentive are rebates. If considering rebates, local governments may base rebates on installation cost or purchasing cost and may limit rebates to installation by local workers. By lowering initial costs, more developers and homeowners are more likely to implement renewable systems.[6] In addition, local governments may seek to help developers and homeowners expedite the permitting of renewable systems. Doing so, may help make it easier for individuals to shift to renewable systems.

Federal and state governments also offer different types of incentives to promote the purchase and use of renewable energy sources.[7] The Federal government has tax credits that can be found in a few places including the IRS website.[8] Other federal and state incentives and regulations regarding renewable energy can be found on DSIRE.org (Database of State Incentives for Renewables & Efficiency).[9]

EFFECTS

There are multiple benefits stemming from ordinances creating incentives for renewable energy, including economic, ecological, and health benefits.[10] The most direct benefit is a decrease in GHG emissions released from the burning of fossil fuels to create electricity.[11] In addition, the renewable energy field is labor intensive and may provide additional jobs. For example, California alone has over 100,000 jobs in the solar industry.[12] Workers are needed to maintain and install solar panels, wind turbines, and others. This can create new local jobs, in employment areas that have been growing. These jobs have a ripple effect on local business's benefiting the entire economy in the area.[13] With the addition of renewable energy sources local energy prices become more stabilized, as inexhaustible energy sources such as wind and solar produce power that is not affected by shifting costs in fossil fuels.[14] Owners of renewable energy systems see immediate savings on their electric bill that can have substantial cost benefits for the owner by exceeding the initial cost of installation.[15] The reduced emissions also lead to improved air quality, benefiting public health.[16]

EXAMPLES

San Francisco, CA

San Francisco offers a series of incentives for residential installation of solar panels. San Francisco offers funds to residents seeking to install solar panels. The rebates go up to $500.00 per kilowatt hour generated by the solar system. The funds can used toward the design, purchase, and installation costs of solar panels.[17] If the panels are installed by an individual, firm, or organization located in the City, an applicant can receive an extra $250.00 per kilo watt hour, towards a solar system.[18] The San Francisco environment code also created an Administrator of the Solar Incentive Program (the Public Utilities Commission) and gave the position the ability to create rules for the allocation of solar incentive funds.[19] Funds are allocated from the San Francisco Public Utility Commission's power revenue, along with any surplus from the solar incentive program from the previous year.[20] With the approval of the Solar Incentive Program Administrator, other forms of individual owned renewable generation systems can be given incentives under the program created by the environment code.[21] San Francisco has no generation capacity for eligible systems, meaning the system can generate more power than is used (see Zero Net Energy Buildings). Incentives are available to only the owner of the renewable energy system.[22] The City's program has been extended to other qualifying renewable energy generation systems.

To view this provision see San Francisco, CA, Environment Code § 18 (2017).

Georgetown, TX

Georgetown offers multiple incentives for residents wanting to add renewable energy sources to their property. One of the incentive programs is net metering. The Georgetown net metering program allows customers of the City's electric utility to connect their solar systems and be credited for excess renewable energy production (i.e., producing more energy than they use) (see Zero Net Energy Buildings).[23] Georgetown also offers different rebates for the installation and purchasing of equipment needed to install solar systems. Rebates are based on panel ratings, inverter ratings, and the efficiency of the system installed.[24] The rebates are available for residential and small commercial customers, for solar, and for other renewable energy generation systems. Georgetown also has specific regulations for solar panels, giving the City more comprehensive coverage for the common renewable energy source.[25] The specific solar provisions make it easier for oversight of solar systems, while not limiting other forms such as wind, geothermal, and biomass.[26]

To view this provision see Georgetown, TX, Code of Ordinances § 13.04.083 (2012).

ENDNOTES

1   U.S. Energy Information Admin., *How much of U.S. carbon dioxide emissions are associated with electricity generation?*, (May 10, 2017), https://perma.cc/P6X4-VQ2V (last visited May 22, 2018).

2   EPA, *State Renewable Energy Resources*, https://perma.cc/Z35K-C5JV (last visited May 25, 2018).

3   Solar Energy Industries Ass'n, *Top 10 Solar States*, (2017), https://perma.cc/Y2A9-TQYV (last visited May 29, 2018).

4   San Francisco, CA, Environment Code §§ 18.1 –18.4 (2017).

5   Georgetown, TX, Code of Ordinances § 13.04.083 (D) (2) (2012).

6   *Barriers to Renewable Energy Technologies*, (Dec. 2017), https://perma.cc/L5MK-W3MD (last visited May 29, 2018).

7   Jerome L. Garciano, Robinson & Cole L.L.P., *Green Tax Incentive Compendium*, Jan. 1, 2018, https://perma.cc/NHY4-K9F3.

8   IRS, *Renewable Electricity, Refined Coal and Indian Coal Production Credit* (2017 tax form), https://perma.cc/U78W-JP6W.

9   N.C. State University, *Database of State Incentives for Renewables & Efficiency*, DOE, https://perma.cc/7QKF-PYTP (last visited May 29, 2018).

10   Union of Concerned Scientists, *Benefits of Renewable Energy Use* (Dec. 20, 2017), https://perma.cc/XY55-XGYP (last visited June 5, 2018).

11   Union of Concerned Scientists, *Clean Power Green Jobs: A National renewable Electricity Standard Will Boost the Economy and Protect the Environment* (Mar. 2009), https://perma.cc/GH26-9782.

12   *See* Solar Energy Industries Ass'n, *supra* note 3.

13   Austin Brown et al., *Estimating Renewable Energy Economic Potential in the United States: Methodology and Initial Results* (NREL 2016), https://perma.cc/FY2Q-SNUQ.

14   Union of Concerned Scientists, *supra* note 10.

15   Hannah West, *Long and Short Term Benefits of Solar Power at Home*, Proud Green Home (July 9, 2015) (noting a savings of $60,000 in 25 years), https://perma.cc/SH9T-JZES (last visited June 1, 2018).

16   EPA, *supra* note 2; N.C. State University, *supra* note 9.

17   San Francisco, CA, Environment Code § 18.4 (b) (1) (2017).

18   *Id.* § 18.4 (b) (5).

19   *Id.* § 18.6.

20   *Id.* § 18.1 (p).

21   *Id.* § 18.3 (b).

22   *Id.* § 18.2 (b).

23   Georgetown, TX, Code of Ordinances § 13.04.083 (D) (2) (2012).

24   *Id.* § 13.04.083 (D) (3) (b).

25   *Id.* § 13.04.083 (G).

26   *Id.* § 13.04.083 (B).

TRANSIT-ORIENTED DEVELOPMENT

Kyler Massner (author)

Jonathan Rosenbloom & Christopher Duerksen (editors)

INTRODUCTION

Transit-oriented developments (TODs) represent a variety of methods and strategies to shape and encourage development around public transportation hubs.[1] TODs seek to leverage the benefits of public transportation in a desire to create compact, walkable, pedestrian-oriented neighborhoods along transit hubs while reducing reliance on automobiles.[2] A TOD ordinance can operate by either creating incentives or requiring particular types of development around transit hubs. Incentive-based ordinances provide a number of benefits to developers, such as density, area or height bonuses, if the development complies with certain requirements.[3] In contrast, zoning codes can also require that developments meet certain requirements before development may begin. Requirements such as mixed-use minimums, density minimums, and maximum limits on available parking, must be achieved if the development is located within a designated TOD district.[4] Both incentive-based and mandatory systems each have benefits.[5] Mandatory programs are easier to control and allow for more predictive outcomes, while incentive-based systems can be broadly applied and sufficiently flexible to meet the needs of a particular circumstance.[6]

EFFECTS

Local governments can implement TOD districts to create areas around public transit stations, which encourage individuals to fulfill the majority, if not all, of their daily needs without the use of automobiles. Any need not immediately accessible within the TOD district should be easily satisfied by use of the transit system, where the need can then be fulfilled by riding to the adjacent TOD district. Since individuals can rely on public transit for their daily commutes and everyday needs, TOD's have the ability to reduce automobile traffic by encouraging individuals to either walk, bike, or use public transportation.[7] Since TOD's reduce the need of automobiles, the result is less traffic and less Greenhouse Gas (GHG) emissions, particularly when compared to development produced by conventional Euclidian zoning.[8] For example, a case study of a TOD in Phoenix, AZ, found that there was a long-term decrease in GHG emissions due to decreased reliance on vehicles as a result of the TOD.[9] Additional studies have found that TODs have a positive impact on public health because of the convenience of the pedestrian lifestyle.[10]

EXAMPLES

Chicago, IL

Chicago began its TOD program in 2013 and expanded it in 2015. The program seeks to encourage developments within a specified radius of a transit hub by offering incentives to developers who choose to build there.[11] Chief among these incentives is the allowance of fewer parking spaces than the typical minimum and an increase in the permitted floor area ratio.[12] The Chicago ordinance allows for a 100% reduction in required parking if the new TOD development is located within 1,320 feet of a rail station.[13] This allows the developer greater flexibility when utilizing the space by removing impediments to development and increasing the area in which developers can choose to use. These types of financial incentives not only provide a benefit to the developer, but also stimulate devel-

ADDITIONAL EXAMPLES

City of Los Angeles, CA, Code of Ordinances §§ 12.22(25) (2008); 13.09(E) (1998); 13.07 (1992) (encouraging affordable housing within TODs by providing developers with height, density, and parking flexibility). Also see Los Angeles' Transit Oriented Communities Affordable Housing Incentive Program Guidelines.

Austin, TX, Ordinance No. 20050519-008 (2005) (establishing different classification of mandatory TOD districts dependent on the district's location and role in the transit system).

San Francisco, CA, Planning Code § 209.4 (2017) (establishing TOD with specific density and parking requirements that are designed to be used near transit stations).

South Miami, FL, Land Development Code §20-8.1 (1997) (establishing TOD limiting drive thru business, requiring 70% onsite parking and a devotion of 75% of first floor wall space to windows aimed at generating pedestrian interest).

Capitol Region Council of Governments, Sustainable Land Use Code Project, Model TOD Ordinance, https://crcog.org/2016/04/sustainable-land-use-regulation-project-crcog-model-land-use-regulations.

opment around transit hubs. By eliminating required parking, space is made available for other mixed uses that can occupy the space, thereby facilitating the purpose of the TOD.

To view the provision see Chicago, IL, Zoning Code §§ 17-3-0403-B (2018); 17-10-0102-B (2018).

Bloomington, MN

Bloomington's zoning code seeks to encourage high intensity mixed-use areas close to transit services. Bloomington's ordinance is notable as it combines elements of both minimum requirements and incentive-based development practices. The ordinance requires that only principle uses that seek to advance a pedestrian orientated lifestyle and increase transit efficiency, such as apartment buildings, hotels, office space, and recreationally-oriented spaces, are permitted in the district.[14] Developers in the TOD district are incentivized to provide high intensity, pedestrian oriented developments by receiving an increase in the maximum floor area ratio if certain criteria are met.[15] For example, developers that include a publicly accessible park can increase the floor area in a 1:1 ratio for each square foot of publicly accessible park.[16] In addition, developments which include retail and service space can qualify for up to a 50% bonus of square footage, while developments that allow for below grade parking can have up to a 75% bonus.[17] Encouraging high intensity development removes barriers to pedestrian traffic by increasing the amount of walkable area within the district. The removal of these barriers not only promotes foot traffic for local business, but also reduces automobile traffic and associated GHG emissions.

To view the provision see Bloomington, MN, Code of Ordinances § 19.29 (2018).

ADDITIONAL RESOURCES

EPA et al., *Encouraging Transit Oriented Development: Case Studies That Work*, https://www.epa.gov/sites/production/files/2014-05/documents/phoenix-sgia-case-studies.pdf.

Suzanne Rhees, *Transit-Oriented Development from Policy to Reality*, crplaning.com, https://perma.cc/W6NQ-PJLD (last visited June 2, 2017).

Urban Land Institute, *Fiscal Impacts of Transit-Oriented Projects*, https://perma.cc/HQ22-ZQUB (last visited June 15, 2017).

CITY OF DES MOINES, COMPREHENSIVE PLAN 13, 28 (2016), https://perma. cc/77UF-7MGZ (setting goals that integrate transit-oriented development in as a priority for both transit and land use planning).

ENDNOTES

1   RAY HUTCHISON, ENCYCLOPEDIA OF URBAN STUDIES 823 (2010); ADRIENNE SCHMITZ & JASON SCULLY, CREATING WALKABLE PLACES: COMPACT MIXED-USE SOLUTIONS 26-28 (2006).

2   HUTCHISON, *supra* note 1, 823

3   *See, e.g.,* Chicago, IL, Zoning Code § 17-10-0102-B (2017); Bloomington, MN, City Code § 19.29 (2018).

4   *See, e.g.,* Los Angeles County, CA, Ordinance No. 2005-0011; Austin, TX, Ordinance No. 20050519-008.

5   HUTCHISON, *supra* note 1, 823; SCHMITZ & SCULLY, *supra* note 1, 26-27.

6   HUTCHISON, *supra* note 1, 823; SCHMITZ & SCULLY, *supra* note 1, 26-27.

7   JAN GEHL, CITIES FOR PEOPLE 107-09 (2010).

8   *Id.*; Mikhail Chester & Dwarakanath Ravikumar, *Transit-Oriented Development Deployment Strategies to Maximize Integrated Transportation and Land Use Life Cycle Greenhouse Gas Reductions*, Proc. ISSST v1 (2013), http://dx.doi.org/10.6084/m9.figshare.805094.

9   Mikhail Chester & Dwarakanath Ravikumar, *Transit-Oriented Development Deployment Strategies to Maximize Integrated Transportation and Land Use Life Cycle Greenhouse Gas Reductions*, Proc. ISSST v1 (2013).

10 SCHMITZ & SCULLY, *supra* note 1, 26-27; John Pucher et al., *Walking and Cycling to Health: a Comparative Analysis of City, State, and International Data*, 100 AM. J. PUB. HEALTH 1986, 1990-1991 (2010).

11 Chicago, IL, Zoning Code §§ 17-3-0403-B, 17-10-0102-B (2017).

12 *Id.*

13 *Id.* § 17-10-0102-B (2017).

14 Bloomington, MN, Code of Ordinances § 19.29(b) (2018).

15 *Id.* §§ 19.29(f)(1), 19.29(g)(4).

16 *Id.* § 19.29(g)(4)(C).

17 *Id.* § 19.29(g)(4)(A)-(B).

Varying Unit Sizes within Multi-Family and Mixed-Use Buildings

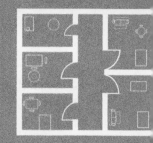

Alec LeSher (author)
Jonathan Rosenbloom & Christopher Duerksen (editors)

INTRODUCTION

Creating a variety of unit sizes (VUS) within multi-family residential and mixed-use buildings is a market-based approach to addressing the need for affordable and sustainable housing development. Many municipalities establish minimum living space requirements, as well as designated living areas, such as kitchens and bathrooms.[1] These restrictions prevent developers from building smaller and more efficient living spaces. In contrast, VUS ordinances require and/or incentivize developers to build a wide variety of unit sizes—from small efficiency apartments up to three-bedroom apartments—within the same multi-family development.[2] Recognizing that housing markets include wide varieties of buyers and renters, local governments can implement VUS ordinances to expand the supply of affordable housing options, and to improve the sustainability of neighborhoods.

EFFECTS

Traditional ordinances and building practices limit the supply of multi-family dwelling units to uniform sizes, forcing most potential renters and buyers to pay for standard-sized units, even if those units are not best suited to an individual or family's needs and financial capacity. This lack of choice has a broader effect on the market, as it artificially increases demand and prices for standard dwelling units.[3] Further, when coupled with housing incentives and subsidies, it drives up the taxpayer costs of such programs in a cyclical fashion. Housing subsidies encourage developers to build larger, more expensive units rather than small, economically efficient units.[4] When developers build larger and more expensive units, the government is forced to increase subsidies to ensure sufficient affordable housing for low-income individuals.[5] VUS ordinances break this cycle by supplying multiple market-priced options suitable to diverse needs including, small units with low costs of entry, standard units, and larger, high-end units.

By integrating small, standard, and large units within a single multi-family development, VUS ordinances improve the sustainability of the entire community. Varied unit sizes allow populations to concentrate in a city center by ensuring single people and families both have suitable housing options, regardless of income or wealth. This concentration of diverse individuals near centers of employment also reduces the need for automobiles.[6] Reducing the use of automobiles, in turn, reduces GHG emissions.[7] Further, by allowing smaller efficiency units, more people have more housing options that meet their needs and developers are not required to build unnecessarily large, expensive, and energy demanding units.[8] VUS ordinances also allow families to opt for a larger apartment rather than a single-family house in the suburbs, which further increases population density, helping to decrease GHG emissions.[9]

VUS ordinances offer the potential to provide renters and first-time buyers a lower-cost point of entry into a desired market. As family sizes or budgets grow, these renters or owners can move into progressively larger dwelling units within their existing community. Similarly, as a family's size or budget shrinks (e.g., retired "empty-nesters"), current residents can downsize their dwelling space without leaving their community.[10] VUS ordinances help result in diverse housing options for both new and existing residents, bringing and retaining a diverse population.

One potential criticism of VUS ordinances is that they may help raise the cost of construction by adding design costs and variability to the construction process.[11] While variation in size may produce variation in price per square foot, this is not determinative of overall construction costs nor of the price supported by a market.[12]

EXAMPLES

Seattle, WA

Seattle supports two classifications of small unit dwellings—Efficient Dwelling Units (EDU) and Congregate Residence (CR). The EDUs are residential units that total at least 220 square feet and include an area for food preparation and a bathroom.[13] The CRs are also residential units, but these allow for smaller sleeping areas with access to communal kitchens and recreational areas.[14] The code provisions permitting EDUs and CRs allow for smaller units while also insuring necessary living spaces for a person's daily needs. Not only does the code allow for more compact and sustainable living, but it also allows for more afford-able and diverse housing units as individuals can rent smaller spaces that cost less. The code in turn helps alleviate some of the income disparities and gentrification that are often present in cities such as Seattle.[15]

To view the provisions see Seattle, WA, Municipal Code §§ 23.42.048 (B), 23.42.049 (2014).

Erie, Colorado

Erie has a required VUS ordinance for the purposes of neighborhood architecture, diversity, and quality. It is generally applicable to all new multi-family buildings in order to increase population density, create a variety of housing options, and foster creativity in development, as opposed to sameness or "cookie cutter" development.[16]

The ordinance requires multi-family construction to meet one of three possible types of variation: (1) "[a] minimum of 50 percent of the total planned dwelling units shall vary in size

from other dwelling units by at least 250 square feet"; (2) "[a] maximum of 50 percent of the total planned dwelling units may have the same number of bedrooms"; or (3) "[a] minimum of 10 percent of the total planned dwelling units shall have at least 3 bedrooms."[17] The Unified Development Code also provides for alternate methods of compliance with these requirements.[18] Developers can propose "alternative equivalent[s]" to the variation requirements, subject to approval by the Community Development Director

(CDD).[19] If the CDD finds that the proposal meets or exceeds the intent, goals, and benefits of the provisions then the proposal will be approved as an exception to the general requirements.[20]

To view the provisions see Town of Erie, CO, Unified Development Code 10.6.7. (2017).

ENDNOTES

1   Miami Lakes, FL, Code of Ordinances § 13-525 (g) (2018); Peshtigo, WI, Code of Ordinances § 60-65 (e) (6) (2009).

2   *See, e.g.,* Bainbridge Island, WA, Bainbridge Island Municipal Code § 2.16.020 (Q) (2016).

3   *See* Edward J. Pinto, *Market-Based Solutions Are the Only Way to Get Home Prices and Rents Back in Line*, Am. Enter. Inst. (July 18, 2016), https://perma.cc/PVE8-Y6VL.

4   *Id.*

5   *Id.*

6   D. Dodman, Urban Form, Greenhouse Gas Emissions and Climate Vulnerability 68 (2009).

7   *Id.*

8   *See, e.g.,* Seattle, WA, Municipal Code § 23.42.048 (2018).

9   *See* Erie, CO, Unified Development Code § 10.6.7 (2017) (requiring a percentage of new apartments to have three bedrooms as part of the stated purpose of reducing dependence on automobiles).

10   *See, e.g.,* Greendale, WI, Village of Greendale Comprehensive Plan: 2010 – 2035, § 6-14, https://perma.cc/83ED-GZLL (last visited May 22, 2018) (establishing a goal of making Greendale a "community where residents can 'age in place.'").

11   *See* Austin, TX, Codes and Ordinances Subcommittee (Aug. 19, 2014), https://perma.cc/8MUF-NG24 (last visited May 24, 2018) (estimating that micro-units cost approximately 1.5 - 3 times more than standard apartments on a per-square-foot basis, but overall costs are twenty percent less than standard apartments).

12   *See, e.g.,* Robert Cassidy, *Multifamily Report: How Building Teams Are Beating the Cost Crunch*, Building Design & Constr. (Jan. 1, 2017), https://perma.cc/C4DR-6SJ2 (last visited May. 24, 2018); Bendix Anderson, *Design Makes a Difference in Affordable Housing*, Multifamily Executive (Dec. 18, 2017) https://perma.cc/ZWF2-V6E7; George Howland Jr., *Market-Based, Mission-Driven Developer Dramatically Lowers Costs of Building Affordable Housing*, Outside City Hall (June 1, 2017), https://perma.cc/DXJ8-M2ZG.

13   Seattle, WA, Municipal Code § 23.42.048 (B) (2018).

14   *Id.* § 23.42.049.

15   Knute Berger, *Looking at Gentrification in Seattle*, Seattle Mag. (May 24,, 2016), https://perma.cc/T8EQ-MEMP.

16   *See* Erie, CO, Unified Development Code § 10.6.7 (d) (1) (B) (v) (2017).

17   *Id.*

18   *Id.* §§ 10.6.7 (C), 10.6.1 (C) (2017).

19   *Id.* § 10.6.1 (C) (4).

20   *Id.* § 10.6.1 (C) (5).

Part 3:

FILL REGULATORY GAPS

ALTERNATIVE PEDESTRIAN ROUTES TO PARKING AREAS, NEIGHBORHOODS, AND BUSINESSES

Kyler Massner (author)

Charlie Cowell, Jonathan Rosenbloom & Brett DuBois (editors)

INTRODUCTION

Local governments seeking to increase pedestrian mobility can help by enacting ordinances that require alternative pedestrian routes to and from parking areas, neighborhoods, and businesses.[1] Alternative pedestrian routes include pathways that allow pedestrians safe passage.[2] Some examples include dedicated bicycle or pedestrian pathways over 8 feet wide that connect to existing paths.[3] At a minimum, these ordinances prioritize pedestrian mobility and require developers to provide pedestrian infrastructure.[4] In addition, these ordinances strongly disfavor culs-de-sac or dead-end streets which are permitted only if the developer provides a "cut through" or "access" easement for pedestrians, or to protect environmental features such as rivers, forest, and habitats.[5]

When a local government seeks to require alternative pedestrian routes in its zoning or subdivision codes, success can be best realized when accompanied by a pedestrian mobility section within a comprehensive plan that identifies current and future pedestrian needs.[6] In conjunction with ordinances, such a plan can be used to develop a complete system of alternative pedestrian pathways with safe crossings, efficient routes, and a variety of pedestrian facilities. With a comprehensive plan in place, local governments can require compliance with the plan for all required pathways/sidewalks on new or redevelopment projects.[7]

Whether or not a local government chooses to develop a pedestrian plan, requirements for alternative pedestrian routes can be done within existing codes as either stand-alone ordinances or in standards for district or overlay zones. These ordinances require all new or redevelopment projects to install pathways which connect to existing pedestrian infrastructure.[8] These ordinances also require developers to provide multiple alternative pedestrian pathway options to areas both on and off site, which are visible, convenient, and direct.[9] Alternatives can include things such as building new sidewalks that connect to existing ones, widening existing sidewalks, or adding bicycle lanes.[10] For making pedestrian connections, these ordinances should be sure to measure walking distance based on the actual walkable surface as opposed to measuring the direct distance "as the crow flies."[11]

These ordinances not only apply to neighborhood areas, but also to business and commercial areas where alternative pedestrian pathways are required in parking lots and between adjacent buildings/properties.[12] For example, where a project contains more than one principal building or are over a certain size, these ordinances require a dedicated pedestrian pathway between principal and accessory buildings, between adjacent properties, and one along every street entrance.[13] When requiring alternative pedestrian routes in high automobile traffic areas such as business and commercial areas, it is important that a local government make proper allowances for pedestrian safety that differ from residential areas, such as, but not limited to, wider pathways, signalized crosswalks, and pathway lighting.[14]

EFFECTS

The US Department of Transportation (DOT) finds that "with the exception of work, the large majority of trips each day are less than five miles" thus presenting a major opportunity to promote walkability within our communities.[15] The DOT also recognizes that increased physical activity is associated with municipalities that have high levels of connectivity (i.e., pedestrian infrastructure characterized by "direct routing, accessibility, few dead-ends, and few physical barriers").[16] Furthermore, the Centers for Disease Control (CDC) finds that expanding the availability of pedestrian transportation options increases the tendency to walk which can save lives by preventing chronic diseases, reducing and preventing motor-vehicle-related injury and deaths, improving environmental health, while stimulating economic development, and ensuring access for all people.[17] By placing pedestrian mobility as a priority, ordinances that require alternative pedestrian pathways can improve community health, reduce greenhouse gases, and create a more equitable community.[18]

By requiring alternative pedestrian pathways, walking distances within a community are shortened, which can lead to trips by foot being faster than driving and encourage individuals to walk.[19] Also, the investment into pedestrian infrastructure works to encourage walking by providing amenities for pedestrians and making a trip by foot safer than by car.[20] Thus, by requiring alternative pedestrian pathways, municipalities can promote healthy choices and encourage the individual to reap the health benefits of walking.[21] Not only is health promoted, but these ordinances also have positive environmental/energy benefits since the more people walking reduces the amount of vehicle miles traveled, thus reducing green-house gases emitted by automobiles.[22] Additionally, the US Census Bureau finds that low-income individuals are most likely to bike and walk to work, while only two-thirds of Americans can afford or are legally permitted to drive a vehicle.[23] These statistics raise equity issues pertaining to mobility. By requiring alternative pedestrian pathways, a municipality can promote/improve pedestrian mobility for all members of the community, leading to more equity in transportation options.[24]

EXAMPLES

Bannock County, ID

Bannock County establishes zoning districts to separate land uses and mitigate the effects of land use conflicts.[25] For the purpose of "creat[ing] self-sustaining new communities with integrated commercial, recreational, natural, and residential land uses," the Master Planned Community (MPC) zoning district requires alternative pedestrian routes in neighborhoods, parking lots, and business areas.[26] Developers in MPC districts are required to consider pedestrian mobility and integrate a pedestrian mobility network that provides multiple pathways to areas within the community.[27] Internal pedestrian pathways which connect to off-site pedestrian infrastructure is a requirement in a MPC district.[28]

To facilitate pedestrian mobility, the County strongly discourages private streets while gated communities are outright prohibited.[29] Developers are also permitted to incorporate a variety of sidewalk designs (i.e., meandering pathways or pedestrian streets) provided that such walkways connect to the larger trail system.[30] Additionally, the County makes sidewalks on both sides of the street the standard, which will only be waived if the developer constructs a dedicated bicycle or pedestrian pathway which provides access to other pathway systems.[31] The County also may require pedestrian amenities (i.e., bulb outs and other pedestrian features) to shorten walking distances and increase pedestrian safety depending on the nature of the block, while crosswalks are required to be incorporated at key intersections, mid-blocks, and parking lots.[32]

To view the provisions see Bannock County, ID, Code of Ordinances § 17.46 (2016).

Atlantic Beach, NC

The city of Atlantic Beach has an "Access and Circulation" section within their Unified Development Ordinance that is applicable to all developments within the municipality.[33] The City describes the purpose of the section as creating a highly connected transportation system that provides mobility choices to its citizens.[34] Specifically, the City sets pedestrian mobility as a top priority and seeks to promote walking by requiring alternative pedestrian pathways between neighborhoods, parks, schools, and commercial areas.[35]

The City executes this mission by requiring sidewalks to be installed along the frontage of all new and redevelopment projects within many residential and commercial districts.[36] The City requires that all development adjacent to vacant land continue all pedestrian and bicycle pathways to the property line of that parcel for continuance of future pedestrian infrastructure.[37] Furthermore, the City strongly discourages culs-de-sac or dead-end streets and when permitted, the developer must

ADDITIONAL EXAMPLES

Marietta, GA, Code of Ordinances § 712.09 (2008) (establishing a district that requires alternative pedestrian routes and mandates sidewalk safety measures, pedestrian facilities, and inter-parcel pedestrian access between adjacent lots and within parking lots).

Longmont, CO, Code of Ordinances § 15.03.150 (2015) (requiring developments in the mixed-use district to provide a multi-modal access and circulation plan that integrates off-site pedestrian/bicycle infrastructure with multi-modal connections, cross walks, traffic mitigation improvements, and pedestrian gateways).

Rocky Mount, NC, Code of Ordinances § 712(d)(4)(e)(15) (2011) (providing city officials with the authority to authorize substitution of a sidewalk with alternate pedestrian walkways (i.e., nature trails, direct access pathways)).

Thurston County, WA, Code of Ordinances § 21.70.100 (1996) (requiring pedestrian pathways to be the shortest distance possible and illustrates acceptable parking lot designs that create a pedestrian friendly environment in all commercial developments).

Melbourne, FL, Code of Ordinances §§ 9.107, 9.112 (2018); Comprehensive Plan § 3.1 (2011) (requiring all sidewalks and other pathways identified in the city's comprehensive plan or "Official Future Bikeway/Sidewalk/Pedway Facility Map" to be provided within a new or redevelopment project).

"include a sidewalk or multi-use path from the dead end or culs-de-sac to the closest local street, collector street, or to a culs-de-sac in an adjoining subdivision."[38] Where a sidewalk is not present or where the existing sidewalk system is not connected to other pedestrian infrastructure, a "multi-use path may be proposed in lieu of sidewalks" which are required to connect areas of interest (i.e. schools, shopping centers, parks, etc.), be visible and easily accessible, marked with destination and directional signage, and eliminate crossing vehicle right-a-ways wherever possible.[39] The City requires these multi-use paths to be located in easements dedicated to "pedestrian and bicycle access by members of the general public", and constructed with durable, low-maintenance materials that are adequately wide to provide clearance for multiple users.[40] Additionally, all streets that do not currently have pedestrian improvements are required to be retrofitted under these standards when redevelopment is sought.[41]

Alternative pathways are also required in parking lots. For developments with on-site parking, the City requires pedestrian walkways that "minimize conflict between pedestrians and traffic at all points of pedestrian access to on-site parking and building entrances."[42] Defined as "On-site Pedestrian Circulation," walkways are required to connect building entrances to adjacent building entrances and sidewalks, provide walkways to all access points or parking spaces that are more than fifty-feet from a buildings entrance or on-site destination, and when a project contains two or more principal building, the developer shall provide a "pedestrian walkway between the primary entrance of each principal building."[43]

To view the provisions see Atlantic Beach, NC, Code of Ordinances § 18.5.2 (2018).

Orange County, FL

Orange County supports alternative pedestrian routes by requiring pedestrian circulation in a wide variety of zoning districts.[44] Within designated districts projects are required to connect to both existing and proposed pedestrian and bicycle pathways which integrate the project to "surrounding streets, external sidewalks, outparcels, and transit stops."[45] The County requires pedestrian walkways to safely provide access between parking lots and building entrances by separating vehicular and pedestrian traffic and by constructing dedicated pedestrian pathways.[46] Direct pedestrian pathways between buildings are also required unless prevented by physical limitation.[47]

Furthermore, separated pedestrian access points are required at a "minimum ratio of one access point for each vehicular access point."[48] Further, separated pedestrian access points must be connected to the larger pedestrian network (i.e., public sidewalks, transit stops, outparcels) and located in a manner that provides the "earliest point of off-site pedestrian walkway contact."[49] The County requires these pedestrian walkways to be designed and constructed to move people safely from building and parking areas by requiring reasonable breaks in parking lots to allow pedestrian access to points of interest.[50]

To view the provisions see Orange County, FL, Code of Ordinances §§ 38-808, 833, 858, 883, (2013).

ADDITIONAL RESOURCES

EPA, *Smart Growth in Small Towns and Rural Communities*, https://perma.cc/L732-ALGK (last visited June 6, 2019).

EPA, *Healthy Places for Healthy People*, https://perma.cc/ULU9-7STH (last visited June 6, 2019).

U.S. Green Building Council, *LEED v4 for Neighborhood Development* (July 2, 2018), https://perma.cc/4BMB-HDV9.

Douglas Shinkle & Anne Teigen, Encouraging Bicycling and Walking: The State Legislative Role (Nat'l Conference of State Leg. Nov. 2008), https://perma.cc/X2PC-7DY2.

Initiative for Healthy Infrastructure, Univ. at Albany, Planning and Policy Models: For Pedestrian and Bicycle Friendly Communities in New York State (Sept. 2007), https://perma.cc/HU6P-6YMK.

Dep't of Transp., The National Bicycling and Walking Study: 15-Year Status Report (May 2010), https://perma.cc/XU2F-KQMT.

Dep't of Transp., *Promoting Connectivity* (Oct. 26, 2015), https://perma.cc/35NM-FRG6.

Ctr. for Disease Control & Prevention, *CDC Recommendations for Improving Health through Transportation Policy* (Feb. 7, 2018), https://perma.cc/BJ32-Z63V.

Div. of Planning, Ky. Transp. Cabinet, Street Connectivity: Zoning and Subdivision Model Ordinance (Mar. 2009), https://perma.cc/7QPY-QFSE.

CALGreen, Dep't of Hous. & Cmty. Dev., Guide to the 2013 California Green Building Standards Code Residential (2013), https://perma.cc/3SPK-R8SN.

ENDNOTES

1   Douglas Shinkle & Anne Teigen, Encouraging Bicycling and Walking: The State Legislative Role vii (Nat'l Conference of State Leg. Nov. 2008), https://perma.cc/X2PC-7DY2.
2   See Atlantic Beach, NC, Code of Ordinances § 18.5.2(C)(3) (2018), Bannock Cty., ID, Code of Ordinances § 17.46.100(G)(2) (2016), Orange Cty., FL, Code of Ordinances §§ 38-808(a), 833(a), 858(a), 883(a), 1704(a), 1748(a) (2013).
3   Bannock Cty., ID, Code of Ordinances § 17.46.100(G)(2).
4   See, e.g., Atlantic Beach, NC, Code of Ordinances § 18.5.2(C)(4)(a)-(b), (F)(1)(a)(ii).
5   Id.
6   Univ. at Albany, Planning and Policy Models for Pedestrian and Bicycle Friendly Communities in New York State 9-10 (Sept. 2007), https://perma.cc/HU6P-6YMK.
7   See, e.g., Melbourne, FL, Code of Ordinances art. VII, § 9.112 (2018).
8   See, e.g., Atlantic Beach, NC, Code of Ordinances § 18.5.2(F).
9   See Melbourne, FL, Comprehensive Plan ch. 3, obj. 3.1 (2011); Div. of Planning, Ky. Transp. Cabinet, Street Connectivity: Zoning and Subdivision Model Ordinance 1-9 (Mar. 2009), https://perma.cc/7QPY-QFSE.
10   Melbourne, FL., Comprehensive Plan ch. 3, obj. 3.1 (2011).
11   CALGreen, Dep't of Hous. And Cmty. Dev., Guide to the 2013 California Green Building Standards Code Residential 77-78 (2013), https://perma.cc/3SPK-R8SN.
12   See, e.g., Atlantic Beach, NC, Code of Ordinances § 18.5.2.
13   See, e.g., id. § 18.5.2(C), (F).
14   See, e.g., id. § 18.5.2(E).
15   City & Cty. of Denver, City and County of Denver Pedestrian Master Plan 2 (Aug. 2004), https://perma.cc/97F6-68SU.
16   Dep't of Transp., Promoting Connectivity (Oct. 26, 2015), https://perma.cc/35NM-FRG6.
17   Ctr. For Disease Control & Prevention, CDC Recommendations for Improving Health Through Transportation Policy 1 (Feb. 7, 2018) https://perma.cc/BJ32-Z63V [hereinafter CDC]; Dep't of Transp., The National Bicycling and Walking Study: 15-Year Status Report 2-3 (May 2010), https://perma.cc/XU2F-KQMT.
18   Dep't of Transp., supra note 16, at 2; CDC, supra note 17, at 1-9.
19   Dep't of Transp., supra note 16, at 2; CDC, supra note 17, at 1-9; Shinkle & Teigen, supra note 1, at 5-6.
20   CDC, supra note 17, at 1-9.
21   Dep't of Transp., supra note 16, at 2; CDC, supra note 17, at 1-9.
22   Dep't of Transp., supra note 16, at 2; CDC, supra note 17, at 1-9.
23   Tanya Snyder, Low-Income Americans Walk and Bike to Work the Most, StreetsBlog (May 8, 2014), https://perma.cc/D69Q-BQX8.
24   Natasha Frost, For the Good of All Humankind, Make Your City More Walkable, Quartz (Oct. 13, 2018) https://perma.cc/FPD3-RHAR.
25   Bannock Cty., ID, Code of Ordinances § 17.08.020 (1998).
26   Id. § 17.46.010 (2016).
27   Id. § 17.46.100(G)(5), (H).
28   Id. § 17.46.060.

29   *Id.* § 17.46.100(G)(1).
30   *Id.* § 17.46.100(G)(2).
31   *Id.*
32   *Id.* § 17.46.100(G)(3)-(4).
33   Atlantic Beach, NC, Code of Ordinances § 18.1.5.
34   *Id.* § 18.5.2(A).
35   *Id.*
36   *Id.* § 18.5.2(F)(1)(a)(i).
37   *Id.* § 18.5.2(C)(3).
38   *Id.* § 18.5.2(C)(4)(a)-(b), (F)(1)(a)(ii).
39   *Id.* § 18.5.2(F)(2).
40   *Id.* § 18.5.2(F)(2)(a)(iii).
41   *Id.* § 18.5.2(C)(5).
42   *Id.* § 18.5.2(F)(3).
43   *Id.*
44   Orange Cty., FL, Code of Ordinances §§ 38-808, 833, 858, 883, 1704, 1748.
45   *Id.* §§ 38-808(a), 833(a), 858(a), 883(a), 1704(a), 1748(a).
46   *Id.*
47   *Id.*
48   *Id.* §§ 38-808(b), 833(b), 858(b), 883(b), 1704(b), 1748(b).
49   *Id.*
50   *Id.* §§ 38-808(c)-(g), 833(c)-(g), 858(c)-(g), 883(c)-(g), 1350, 1748(c)-(g).

Energy Benchmarking, Auditing, and Upgrading

Tyler Adams (author)
Jonathan Rosenbloom & Christopher Duerksen (editors)

INTRODUCTION

Energy efficiency initiatives simultaneously help reduce energy costs and greenhouse gas (GHG) emissions, while creating a more sustainable building stock. Municipalities use energy benchmarking, auditing, and upgrade requirements in order to encourage property owners to improve buildings in accordance with local sustainability goals. Benchmarking allows prospective and current owners to compare the energy use of various buildings of similar size.[1] Pursuant to these ordinances, owners track their buildings' energy usage by entering energy use data on a monthly basis into tracking tools, such as the Environmental Protection Agency's Energy Star Portfolio Manager. As part of these benchmarking ordinances, municipalities require building owners to annually report a buildings' energy use data either directly to the responsible local agency or, more commonly, through the Portfolio Manager tool. In addition, most jurisdictions require disclosure of the benchmarking reports, making them available to the public.

Local governments may also require energy audits, sometimes called assessments. Audits require a more extensive analysis of a buildings' energy use. Audits also require a third-party to perform the audit. A qualified third-party auditor locates the sources of inefficient energy use, which allows owners to identify the measures that can be taken in order to optimize efficiency.[2] There are different levels of comprehensiveness for audits. Municipalities have the option to require audits to meet certain levels and/or include certain criteria that are important for the particular community.[3]

Jurisdictions that require energy audits typically require them to be completed at least once every five years or on the occurrence of a major real estate event, such as a sale, lease, or major renovation. Based on the information gained from energy benchmarking and auditing, municipalities can require buildings to take further steps to become more energy efficient.

EFFECTS

One of the most common ways to generate energy is by burning fossil fuels, such as oil, natural gas, and coal.[4] This burning process releases potent GHGs and other air pollutants.[5] Commercial buildings are responsible for 19% of the total energy use in the U.S. Increasing the efficiency of buildings would reduce the amount of fossil fuels burned, subsequently decreasing overall GHG emissions.[6] Benchmarking "allows owners and occupants to understand their building's relative performance, and helps identify opportunities to cut energy waste."[7] Through continuous measuring of a building's energy use, owners are able to identify where they are lacking in comparison to other buildings. This drives innovation and can create a more competitive job market with building owners striving to be the most efficient.[8] In addition, a study done by the Environmental Protection Agency indicated that on average buildings that are benchmarked use 2.4% less energy than those that are not, presumably because of increased awareness of energy use.[9]

Requiring periodic energy audits results in further benefits by providing owners with an in-depth analysis on where and how a building is inefficiently using energy. Auditors typically provide a report on what actions can be taken in order to optimize energy use. By following these recommendations, as well as the information gained from benchmarking, building owners can greatly reduce their buildings' energy use. This can result in lowered utility costs, and by achieving Energy Star certification or a high Energy Star score through the benchmarking process, buildings can become more profitable when sold or leased.[10]

EXAMPLES

Atlanta, GA

The city of Atlanta requires both energy benchmarking and auditing. A covered building for benchmarking purposes is either a City-owned building exceeding 10,000 square feet or a non-city owned building classified as Commercial, Exempt, Preferential, or Conservation Use, and exceeding 25,000 square feet.[11] The ordinance requires owners to benchmark and provide annual reports using the EPA's Portfolio Manager tool to the responsible City department.[12] The department makes available to the public the benchmarking information of covered properties with energy performances equal to or greater than an Energy Star score of 55.[13]

A covered building for auditing purposes includes City and non-City owned buildings (of the previously mentioned classifications) that have an area greater than 25,000 square feet.[14] An energy audit must be performed once every ten years and a summary audit report submitted.[15] An energy audit is not required to be performed if the building has achieved Energy Star Certification for at least two of the three years preceding the due date for the audit report or their Energy Star score improved at least 15 points.[16]

To view the provision see Atlanta, GA, Code of Ordinances §§ 8-2222-8-2228 (2016).

Orlando, FL

Orlando enacted its benchmarking ordinance to make accessible comparable building energy usage data, to reduce air pollutant and GHG emissions from energy consumption, to encourage efficient use of energy and water resources, and to promote additional investment in the real estate marketplace.[17] By May 1, 2018 all covered properties, City-owned properties exceeding 10,000 square feet and non-city owned properties exceeding 50,000 square feet, were required to benchmark energy use using EPA's Energy Star Portfolio Manager.[18] The benchmarking has to be performed by a "Qualified Benchmarker" and the property owner is required to annually submit a report through the Portfolio Manager to the director of the City's Office of Sustainability and Energy.[19] The director then makes public all the shared benchmarking information.[20]

ADDITIONAL EXAMPLES

Denver, CO, Code of Ordinances § 4-53 (2016) (annual benchmarking and reporting is required for commercial buildings with a gross floor area of 25,000 square feet or larger).

Seattle, WA, Municipal Code § 22.920.010 (2010) (non-residential buildings larger than 20,000 square feet are required to benchmark and provide annual reports to the City as well as upon request from current and prospective tenants and potential buyers or lenders).

Austin, TX, Code of Ordinances § 6-7-31 (2011) (owners of commercial facilities 10,000 square feet or larger are required to annually calculate an energy use rating for the facility using an audit or approved rating system).

Boulder, CO, Code of Ordinances § 10-7.7-3 (2016) (covered commercial buildings are required to annually rate and report their energy use in a manner prescribed by the city manager and perform an energy assessment within the first three years after the first report, and at least once every ten years thereafter).

Starting December 1, 2020, and annually thereafter, owners of covered properties that receive Energy Star benchmark scores lower than fifty will be notified and then required to perform an energy audit or retro-commissioning.[21] Retro-commissioning, according to the ordinance, means "a systematic process for optimizing the energy efficiency of existing base building systems through the identification and correction of deficiencies in such systems."[22] A summary audit or retro-commissioning report is then required to be filed by May 1, 2025, and once every five years thereafter.[23] The ordinance does not require an owner to perform an audit or retro-commissioning if the building achieves an Energy Star score of 50 or higher or improves its score by ten points.[24] If a building satisfies either criteria within that five-year grace period, they will not have to submit a summary report.

To view the provision see Orlando, FL, Code of Ordinances § 15.03, 15.08 (2016).

ADDITIONAL RESOURCES

Institute for Market Transformation, *Jurisdictions*, Building Rating, http://perma.cc/5M2R-Y4JK (last visited June 14, 2018).

Energy Disclosure Compliance, WegoWise, http://perma.cc/4HBW-74RT (last visited June 14, 2018).

Institute for Market Transformation, *Comparison of U.S. Commercial Building Energy Benchmarking and Transparency Policies*, U.S. Commercial Building Policy Comparison Matrix (Apr. 23, 2018), https://perma.cc/CL29-5HPK.

ENDNOTES

1 Sara Mattern, Note, *Municipal Energy Benchmarking Legislation for Commercial Buildings: You Can't Manage What You Don't Measure*, 40 B.C. Envtl. Aff. L. Rev. 487, 488 (2013).

2 Tony Liou, *What is a Commercial Building Energy Audit?*, GlobeSt.com (Oct. 20, 2011), https://perma.cc/3L58-KJFV.

3 *See generally ASHRAE Audit Level 1 2 and 3: What's the Difference?*, SmartWatt (Jan. 10, 2017),https://perma.cc/F2A8-ZPSU.

4 *Energy Use, The Source of Most Carbon Emissions*, Climate Communication, http://perma.cc/Y5A8-HSKJ (last visited June 18, 2018).

5 *Id.*

6 Mattern, *supra* note 1, at 490.

7 *Energy Benchmarking and Transparency Benefits*, Institute for Market Transformation, http://perma.cc/CMB8-FC3N (last visited June 14, 2018).

8 *See id.*

9 *The Benefits of Energy Benchmarking*, SmartWatt (Oct. 31, 2016), http://perma.cc/5JZP-948L; *see also* Zachary Hart, The Benefits of Benchmarking Building Performance 8 (Institute for Market Transformation Dec. 2015), https://perma.cc/STL3-ZPFK.

10   *See* HART, *supra* note 9, at 9-12.
11   Atlanta, GA, Code of Ordinances § 8-2002 (2016).
12   *Id.* § 8-2222.
13   *Id.* § 8-2226.
14   *Id.* § 8-2002.
15   *Id.* § 8-2227.
16   *Id.*
17   Orlando, FL, Code of Ordinances § 15.01 (2016).
18   *Id.* § 15-03.
19   *Id.*
20   *Id.* § 15.05.
21   *Id.* § 15.10.
22   *Id.* § 15.02.
23   *Id.*
24   *Id.* § 15.08.

GREEN ZONES

Brandon Hanson (author)
Jonathan Rosenbloom & Christopher Duerksen (editors)

Green zones (also known as "ecodistricts") are stationary or floating districts created by a local government to promote sustainable practices, to help reduce environmental impacts, and to help revitalize an area. Green zones are areas that provide local governments with the flexibility to focus on a variety of issues related to sustainability. For example, a local government may use green zones to help promote healthy lifestyles, reduce pollution, and/or provide affordable housing and sustainable jobs. Green zones can be created in a variety of ways, including zoning a specific area as a stationary "Green Zone" or green zones can be drafted to create floating zones, whereby a neighborhood can petition to adopt the floating zone.

Local governments can use multiple strategies within green zones to help reduce pollution. For example, local governments may give higher scrutiny for proposed sources of pollution in the green zone and may give priority permits for programs with designations from sustainable regulatory agencies or third party certifications, such as LEED or Living Building Challenge (see Third-Party Certification Requirements).[1] When a local government is creating a green zone some common provisions include signage to deter diesel truck idling, buffer zones for auto related operations from houses, land use restrictions and others.[2] Many of the briefs in this chapter and other chapters may be incorporated directly into green zones. Local governments should draft these ordinances in a way that helps create a healthy neighborhood, remove/reduce existing environmental concerns, develop green economic opportunities, and encourage community involvement.[3]

Green zones can also be a platform to determine if a regulation will be beneficial for the city though a pilot project. Green zones can serve as a pilot to test new sustainable strategies.[4] When a local government creates a green zone, different stages of implementation can allow resources to be added where needed, permitting the government to assess the needs of the zone to benefit the new green zone.[5] Green zones should be implemented in a way that allows for the local government to monitor their effect. A government can utilize phases to fully understand the impact. Phases may include a formation phase, assessment phase, development phase, and management phase.[6] Green zones may be utilized for a variety of zoning types, including residential and commercial.

EFFECTS

Because green zones are a flexible zoning tool that can help local governments meet a variety of challenges, the possible effects from green zones can be vast and varied based on the local community's desires and needs. Green zones, for example, can help reduce greenhouse gas (GHG) emissions, mitigate damage to water sheds, create jobs, provide for healthier lifestyles, reduce waste, and revitalize districts or areas that need the most help.[7] Green zones can also increase office and residential floor space while cutting GHG emissions.[8] Green zones can also help promote equality by providing job creation and investment opportunities in socially and economically diverse areas which can help mitigate displaced workers.[9] Moreover, green zones can provide access to recreational activities and local healthy foods or improve air quality, depending on what the local government chooses to address. Finally, green zones can be designed to focus solely on energy and/or water needs to help reduce GHG emissions, for example by incorporating zero net energy requirements (see Zero Net Energy Buildings for ordinances promoting net zero buildings)[10] or incorporating permeable surfaces (see Pervious Cover Minimums and Incentives) and other water conservation practices.

EXAMPLES

Los Angeles, CA

Los Angeles creates Clean Up Green Up (CUGU) districts to reduce health impacts from land use issues.[11] CUGU districts require approval for certain types of properties before adding altering existing structures or building new ones. Also, appropriate signage for vehicles, such as "no idling" is required in areas that commonly experience idling. The City also has different building height requirements, distancing requirements (buffer zones), and surface material requirements (permeable surfaces and parking lot requirements) in the CUGU.[12]

Los Angeles also has three ecodistricts: Boyle Heights, Pacoima/Sun Valley, and Wilmington.[13] Because these areas have been designated as some of the most vulnerable for environmental problems by the State, the City chose them for the ecodistricts. In the ecodistrict, some of the requirements include one tree for every four parking spaces in a parking lot, and landscape requirements for yards, which include designating specific types of trees and bushes. Along with green zones, the City is implementing City-wide ordinances to compliment the newly formed zones, such as oil refinery Safeguards, and mandatory air filters on building within 1000 feet of a freeway.[14]

ADDITIONAL EXAMPLES

Washington, DC, SW Ecodistrict Plan (Commission Action) (2013) (Washington D.C. has a comprehensive plan to invigorate an area just south of the national mall that has been underutilized and in need of a revitalization. The plan includes permeable surface materials, tree canopy regulations, and other environmental and socioeconomic improvements).

Alexandria, VA, *Old Town North Small Area Plan: Eco-District* (2015) (Alexandria developed a comprehensive plan for an area with multiple aspects of sustainability and is planning on tracking the area over time to assess the impact of the plan).

Portland, OR, *Lloyd Ecodistrict Roadmap* (Nov. 2012) (Portland developed a district that has building efficiency requirements for existing and new buildings in the area, along with aggregated renewable energy programs and other green infrastructure standards).

To view this provision see Los Angeles, CA, Municipal Code §13.18 (2016).

Minneapolis, MN

Minneapolis by resolution created its green zone initiative after adopting the Minneapolis Climate Action Plan.[15] The City created a Green Zone working group to develop criteria to determine which areas should be designated as Green Zones. The group used a variety of factor to determine which communities would qualify to be designated as a green zone.[16] The factors used for this determination include economic and health disparities, pollution impact, and adverse effects of climate change.[17] The resolution lists goals, including soil clean up, improved air quality, increase "green" jobs, and others. The group that decides what will be done to reach these goals is comprised of 11 community member and four city staff, who recommend to the City Council a plan with multiple actions to be done, which must then be approved by the Council.[18] The City is requiring local agencies to supply support to ensure the green zones achieve their goals, as well as advance those goals.[19]

Some of the plans to help achieve the goals in the resolution are incentives, such as lower tax rates, for businesses to reduce emissions.[20] Actions to overcome barriers preventing green jobs in the area are also being implemented, providing transportation and connections to jobs and employers with in the green zone.[21] The City also allows for community gardening in parks, helping to make healthier food options more accessible.[22] The working group also created plans to engage schools in the green zone to educate children on the environment even when not in the green zone.[23]

To view this resolution see Gordon, Cano & Reich, *Establishing Green Zones in the City of Minneapolis,* Resolution (2017).

ENDNOTES

1   California Environmental Justice Alliance, *Green Zones for Economic and Environmental Sustainability: A Concept Paper from the California Environmental Justice Alliance,* https://perma.cc/DNE2-WFMF (last visited June 12, 2018).

2   LA Collaborative for Environmental Health & Justice, Clean Up Green Up (Aug. 2015), https://perma.cc/MDN6-S94S.

3   California Environmental Justice Alliance, *supra* note 1.

4   Center for Earth, Energy & Democracy, *CEED Fact Sheet: Green Zones,* https://perma.cc/SXJ9-A2QD (last visited June 18, 2018).

5   National Capital Planning Commission, The SW Ecodistrict: Programmatic Design Concept Summary 10th, SW and Interim Banneker (March 2015) https://perma.cc/9SPD-MMD6.

6   Alexandria, VA, Old Town North Small Area Plan: Eco-District (2015), https://perma.cc/VZT6-L9YC.

7   Minneapolis Sustainability Office, Minneapolis Climate Action Plan: A Roadmap to Reducing Citywide Greenhouse Gas Emissions (June 28, 2013), https://perma.cc/NS4H-K24S.

8   Rebecca Sheir, *How D.C. is Turning a 'Pedestrian Dead-Zone' into an Eco-Showcase,* Metro Connection (April 17, 2015), https://perma.cc/KT6H-REXV.

9   EcoDistricts, The EcoDistricts Framework: Building Blocks of Sustainable Cities (May 2013), https://perma.cc/JCN3-TDAF.

10   *Id.*

11   Los Angeles, CA, Municipal Code §13.18 (A) (2016).

12   *Id.* §13.18 (F) (2).

13   LA Collaborative for Environmental Health & Justice, *supra* note 2.

14   *Id.*

15   Gordon, Cano & Reich, *Establishing Green Zones in the City of Minneapolis,* Resolution (2017).

16   *Id.*

17   *Id.*

18   *Green Zones Frequently Asked Questions (FAQ),* https://perma.cc/GV8N-2L22 (last visited June 28, 2018).

19   Gordon et al., *supra* note 15.

20   *City of Minneapolis: Green Zones Initiative Presentation of Recommendations,* April 17, 2017, https://perma.cc/G39K-MRRB.

21   *Id.*

22   *Id.*

23   *Id.*

Limit Solar Restrictions in HOAs and/or CC&Rs

Caragh McMaster, Tegan Jarchow (authors)
Darcie White, Sara Bronin, & Jonathan Rosenbloom (editors)

INTRODUCTION

Covenants, Conditions, and Restrictions (CC&Rs) are private-sector regulations governing what homeowners can do with their land.[1] An important characteristic of CC&Rs is that they run with the land, meaning they stay in place upon sale of a property.[2] CC&Rs are often utilized by Homeowners Associations (HOAs), which are often the outcrop of subdivided properties.[3] CC&Rs outline the rights and obligations among homeowners and between homeowners and HOAs. They often restrict uses, require insurance, and describe other legal obligations.[4] In doing some, some CC&Rs include restrictions prohibiting or severely limiting solar panel installation.[5] These CC&Rs are justified on the grounds that they promote uniformity and aesthetic standards within the community.[6] While it is possible for CC&Rs to be amended, this often requires a supermajority, making change difficult, at times.[7]

With the increased push for green energy, the possibility of prohibiting CC&R solar restrictions is becoming a viable option.[8] In 2010, the State of California amended its Solar Rights Act of 1978 stating "[a]ny covenant, restriction, or condition . . . that effectively prohibits or restricts the installation or use of a solar energy system is void and unenforceable."[9] Despite forty states adopting solar access statutes, only twenty-one of those states expressly mention CC&Rs similar to the California statute.[10] Proponents of solar rights in other states have had to rely on local governments to prohibit CC&R solar restrictions.[11]

Ordinances prohibiting CC&R restrictions should define the scope of the ordinance.[12] The scope may vary and may include preventing the CC&Rs from banning solar installation outright,[13] proscribing "unduly restrictive" CC&Rs, and allowing CC&R restrictions so long as they are "reasonable restrictions" on solar installation. Local governments should be conscious that some language may leave significant discretion to HOAs and courts to interpret the provisions.[14]

Local ordinances may seek to narrow the standard or explicitly define examples of reasonable restrictions.[15] They may also include prohibiting any CC&Rs from having a negative impact on the cost or performance of a solar system.[16] CC&Rs that have such an impact may include restrictions on the system's size, orientation, and tilt,[17] and restrictions on aesthetic grounds.[18] Some local governments have explicitly stated their intent in passing the ordinances as preserving private solar energy production, so as to help guide judicial interpretation of the ordinance.[19] Other local governments have prohibited HOAs from taking too much time in issuing a permit for a homeowner's solar installation. Such ordinances provide a time frame for approval.[20]

EFFECTS

Homeowners will only invest in the expense of solar panels if they have "legally recognized protection to install a solar system . . ."[21] Prohibiting CC&Rs from restricting solar energy systems would give homeowners surety and encourage solar development. Notwithstanding many states' laws protecting homeowners from CC&Rs prohibiting solar installation, courts generally rely on a "reasonableness" test to settle disputes over CC&R provisions, which often defer to HOA interpretations.[22] Such a standard is relatively low and provides homeowners with minimum security. Ordinances prohibiting CC&Rs can encourage homeowners to move forward with solar installations and help mitigate greenhouse gas emissions.[23]

Expanding solar energy systems can have a significant environmental impact. More than 25 million housing units are governed by HOAs with over half of them having useable residential solar energy systems.[24] A five percent increase of those housing units without solar could generate 3.3 gigawatts of clean energy, roughly equivalent to taking 1.1 million vehicles off of the roads.[25]

EXAMPLES

Walker, MN

Walker, MN addresses CC&Rs that block solar energy systems by providing that "[n]o homeowners' agreement, covenant, common interest community, or other contract between multiple property owners within a subdivision of Walker shall restrict or limit solar systems to a greater extent than Walker's renewable energy ordinance."[26] The ordinance states that the City encourages the protection of solar rights in new subdivisions, and that existing solar energy system installations are to be protected, consistent with Minnesota state law.[27]

To view the provision see Walker, MN, Code of Ordinances § 109-254(d) (2016).

Laramie, WY

Laramie, WY's code of ordinances states that "[s]olar energy systems shall be a permitted use in all zoning districts" subject to certain requirements outlined later in the code. Most relevant to CC&Rs, the code states that "[p]rivate restrictions on solar energy systems, such as homeowner's association covenants or restrictions, shall not be permitted."[28]

To view the provision see Laramie, WY, Code of Ordinances § 15.14.030(A)(1)(b) (2017).

ADDITIONAL EXAMPLES

Kingsville, TX, Code of Ordinances § 15-6-180(A) (2014) (providing that no HOA, CC&R, or contract between multiple property owners within a subdivision shall restrict the installation of solar systems more stringently than the City's codified solar regulations).

Oglesby, IL, Code of Ordinances § 14.03.020(J)(6) (current through 2019) (providing that no HOA, CC&R or contract between property owners within a subdivision shall prohibit or restrict solar energy system installation; further noting that no HOA energy policy statement shall be more restrictive than Oglesby's energy standards).

East Peoria, IL, Code of Ordinances § 4-19-9(h) (2018) (preventing CC&Rs, HOAs or private contracts between "multiple property owners within a subdivision" from prohibiting or restricting solar energy system installation).

Rowlett, TX, Code of Ordinances § 77-605(H)(4) (2009) (protecting both solar energy systems and wind energy systems from being prohibited by any HOAs, CC&Rs, or "other deed restriction applicable within residential subdivisions").

San Rafael, CA, Code of Ordinances § 14.16.307(A)(4) (2014) (providing that CC&Rs "cannot prohibit installation of solar equipment on buildings").

ENDNOTES

1   THOMAS STARRS ET AL., BRINGING SOLAR ENERGY TO THE PLANNED COMMUNITY: A HANDBOOK ON ROOFTOP SOLAR SYSTEMS AND PRIVATE LAND USE RESTRICTIONS 12, https://perma.cc/F4K6-VZC9 (last visited Jun. 12, 2019).
2   Id.
3   Evan J. Rosenthal, Letting the Sunshine in: Protecting Residential Access to Solar Energy in Common Interest Developments, 40 FLA. ST. U. L. REV. 995, 1006 (2013).
4   See Starrs et al., supra note 1, at 12.

5   SEIA, *Solar Access Rights*, https://perma.cc/H2Z2-JLGT (last visited Jun. 13, 2018).

6   *Id.*

7   Rosenthal, *supra* note 3, at 1024.

8   SEIA, *supra* note 5.

9   Cal. Civ. Code § 714 (2019), https://perma.cc/U725-DAYX (clarifying that "governing documents" means the declaration and any other documents, such as bylaws, operating rules, articles of incorporation, or articles of association, which govern the operation of a common interest development or association).

10   Rosenthal, *supra* note 3, at 1006.

11   *See id.*

12   *Id.* at 1019.

13   BRIAN ROSS, SOLAR ENERGY STANDARDS - URBAN COMMUNITIES: UPDATED MODEL ORDINANCES FOR SUSTAINABLE DEVELOPMENT 9 (Feb., 2014), https://perma.cc/YQM4-SKA3.

14   *Id.* at 1018-19

15   *Id.* at 1019.

16   THE SOLAR FOUND., A BEAUTIFUL DAY IN THE NEIGHBORHOOD: ENCOURAGING SOLAR DEVELOPMENT THROUGH COMMUNITY ASSOCIATION POLICIES AND PRACTICES 13, https://perma.cc/5SE4-T9HV (last visited Jun. 12, 2019).

17   *Id.* at 14-16.

18   *Id.* at 1020.

19   *Id.* at 1021.

20   THE SOLAR FOUND., *supra* note 16, at 9.

21   Joshua M. Duke & Benjamin Attia, *Negotiated Solar Rights Conflict Resolution: A Comparative Institutional Analysis of Public and Private Processes*, 22 J. ENVTL. & SUSTAINABILITY L. 1, 2 (2015), https://perma.cc/ZB28-4U2J.

22   Rosenthal, *supra* note 3, at 999.

23   COLLEEN MCCANN KETTLES, A COMPREHENSIVE REVIEW OF SOLAR ACCESS LAW IN THE UNITED STATES SUGGESTED STANDARDS FOR A MODEL STATUE AND ORDINANCE, SOLAR AM. BD. FOR CODES AND STANDARDS 4 (Oct., 2008), https://perma.cc/4642-7ZVT.

24   THE SOLAR FOUND., *supra* note 16, at 3.

25   *Id.*

26   *See* Walker, MN, Code of Ordinances § 109-254(d) (2016).

27   *Id.* § 109-254(e).

28   Laramie, WY, Code of Ordinances § 15.14.030(1)(b) (2017).

Maximum Size of Single-Family Residences

MAX.

Alec LeSher (author)

Jonathan Rosenbloom & Christopher Duerksen (editors)

INTRODUCTION

The average house size in the U.S. has increased by more than seven hundred square feet since 1973.[1] While large homes may be beneficial to or desired by individual owners, they shift costs to the public and local government. Older, smaller homes may be demolished and replaced with larger homes that disturb the character of the neighborhood.[2] Large homes also produce more greenhouse gas (GHG) emissions, which contribute to climate change.[3] Municipalities can implement ordinances that set a limit on the size of single-family homes to mitigate these harmful effects. Typically, these ordinances seek to limit the spread of "McMansions." McMansions are large houses in suburban neighborhoods that are regarded as oversized in relation to the character of the neighborhood.[4]

Ordinances setting a maximum house size typically regulate the maximum floor area ratio (FAR), which is the portion of the lot that may be covered by a structure. A municipality may also limit the maximum height, number of stories, or total square feet of the house. Setback and minimum yard requirements can also be used to limit how much of a lot may be covered by the house. Some municipalities have varied height restrictions within the lot. For instance, a two-story structure may be allowed at the rear of the lot, whereas only single-story buildings can be constructed near the public right of way.

A municipality can further tailor the requirements based on individual neighborhoods, rather than residential zones as a whole. For example, if a neighborhood has always had larger homes, new large homes could continue to be permitted. However, neighborhoods with smaller homes could have a more restrictive size limit that would help retain the character of the neighborhood and provide many of the economic and environmental benefits discussed below. In this way, developers can still replace old homes, but only if the new home does not increase the impact on the community and environment.

One potential criticism of these ordinances is that they keep certain people and uses out of certain neighborhoods. However, these ordinances are limiting *maximum* home size, as opposed to minimum home size, which can make some areas exclusive. Limiting maximum home sizes prevents intrusively large homes in neighborhoods that are occupied by more moderate homes and by people with more moderate means, thereby protecting middle and lower income property owners.

EFFECTS

An ordinance limiting the maximum size of single-family homes may have a dramatic impact on a municipality's GHG emissions. Residential homes are responsible for GHG emissions related to the demand for heating, cooling, electricity, and water supply, among other things.[5] These demands are met by burning fossil fuels to produce and transport utilities to the house.[6] On average, a typical 2,598 square foot house is responsible for about twenty eight thousand pounds of carbon dioxide emissions each year.[7] In contrast, a "tiny home" of 186 square feet requires an average of two thousand pounds of carbon dioxide each year.[8] As house size increases so too does the output of GHG emissions. If a municipality limits how large houses can be, it also limits how much GHGs are emitted.

Larger homes also require more energy and materials to construct than more moderately sized homes.[9] Construction produces GHGs in four areas: "manufacture and transportation of building materials; energy consumption of construction equipment; energy consumption for processing materials; and disposal of construction waste."[10] Limiting the maximum size of single-family homes may help reduce the construction sector's impact on climate change by reducing GHGs in any and all four of these areas.

This ordinance can also help insure the preservation of historic districts and increase affordable housing options. Often builders merge two adjacent properties and demolish the existing structures in favor of one new, much larger structure.[11] In historic districts, this process replaces culturally valuable old homes with large, new homes that disturb the character of the district. In other residential districts, smaller, more affordable homes are replaced with larger homes that low-income populations cannot afford. Municipalities should keep these effects in mind when considering ordinances that limit house sizes.

EXAMPLES

Los Angeles, CA

In 2017, Los Angeles revised its zoning ordinance and FAR calculations to combat the rise of McMansions.[12] The 2017 ordinance changed the way that the FAR is calculated in residential zones. Any portion of a building with a ceiling height of fourteen feet or higher counts as twice the square footage of that area in the FAR calculation.[13] This has the effect of encouraging owners to build houses with more usable space, or in the alternative, sacrifice overall size in favor of high ceilings. The City also now regulates the size of new residential buildings based on the character of existing houses in residential zones. For example, one residential zone allows for a larger building mass only at the front of the lot, while another allows for larger mass only at the rear of the lot.[14] These ordinances insure that an owner will not buy an older, smaller house on a less expensive lot with the goal of demolishing that house and building a McMansion that occupies nearly the whole lot. Combined, these ordinances help limit the availability of McMansions and their impact on neighborhoods and the environment.

To view the provisions see Los Angeles, CA, Municipal Code § 12.08 (C) (5) (2017).

Austin, TX

Austin, similar to Los Angeles, faced an issue with the spread of McMansions. In order to protect its citizens and neighborhoods from more McMansion-like development, the City implemented "Residential

ADDITIONAL EXAMPLES

Ashland, OR, Land Use Ordinance § 18.2.5.070 (current through Nov. 2017) (limiting new residences in historic districts to a height of thirty feet and restricting FAR ratios to minimize development impact).

Newport, RI, Code of Ordinances § 17.020.050 (2000) (preventing buildings from covering more than twenty percent of a lot).

Santa Monica, CA, Municipal Code § 9.07.030 (B) (2018), Ord. No. 2572CCS (2018) (limiting lot coverage of a two-story house to thirty five percent for the first story, and twenty percent for the second story).

Palo Alto, CA, Municipal Code § 18.12.040 (2007) (limiting maximum house size to 6,000 square feet).

Atherton, CA, Municipal Code §§ 17.32.040, 17.33.040 (current through Sept. 20, 2017) (setting the maximum dwelling size at eighteen percent of the lot size for lots over one acre in size).

Design and Compatibility Standards," which regulate the size of houses in certain districts of the City.[15] New developments in these areas are subject to a FAR ratio of 0.4 square feet of building to 1.0 square feet of lot size.[16] Alternatively, if that calculation returns a FAR that would only allow a building smaller than 2,300 square feet to be constructed, then the City does not apply the .4 / 1.0 FAR and allows the developer to construct a 2,300 square foot building.[17] Further, the City limits building height to a maximum of thirty-two feet.[18]

The ordinances also set forth unique setback requirements that insure no new building can be substantially larger than the others. For the front yard, the new building must be setback as far as other provisions of the Code allow, or alternatively, as far as the average setback of at least four other buildings on the same side of the street.[19] This promotes uniformity in the aesthetic of the neighborhood and prevents new buildings from occupying the entire lot with a McMansion-like structure. The City also establishes "setback planes."[20] These planes are a line beyond which no structure may extend. In general, a line extends 15 feet straight up from lot line and then slants toward the center of the property at a 45-degree angle.[21] This prevents new buildings from encroaching on neighboring properties even if they meet the setback requirements at ground level.

To view the provisions see Austin, TX, Code of Ordinances, tit. 25, subchapter F, §§ 2.1-2.7 (2006).

ENDNOTES

1   United States Census Bureau, *Median and Average Square Feet of Floor Area in New Single-Family Houses Completed by Location*, https://perma.cc/QA5P-64PK (last visited June 6, 2018).
2   Editorial, *Interim McMansion Law is a Fit Addition for Some Areas*, L.A. TIMES, Mar. 17, 2018, at https://perma.cc/NR67-88UL.
3   Gabriella Morrison, *Why Tiny Houses Can Save the Earth Infographic*, TINYHOUSEBUILD.COM (Oct. 26, 2014), https://perma.cc/KHG2-EPHX.
4   *McMansion*, Merriam-Webster, https://perma.cc/LLA5-GETQ.
5   *Sources of Greenhouse Gas Emisssions: Commercial and Residential Sector Emissions*, U.S. EPA (April 11, 2018), https://perma.cc/3QRL-WQAP.
6   *Id.*
7   Morrison, *supra* note 3.
8   *Id.*
9   *See id.*
10   Hui Yan et al., *Greenhouse Gas Emissions in Building Construction: A Case Study of One Peking in Hong Kong*, 45 BLDG. & ENV'T 4, 949 (2010).
11   Editorial, *supra* note 2.
12   Elijah Chiland, *LA Takes New Steps to Fight McMansions*, CURBED LOS ANGELES (Mar. 1, 2017), https://perma.cc/9V9V-U3PF.
13   Los Angeles, CA, Municipal Code §12.03 (2017) (defining "Floor Area, Residential").
14   *Id.* §§ 12.08.5.c, 12.08.5.d.

15   *Residential Design and Compatibility Standards*, Development Services Department , https://perma. cc/2CBE-LNNM (last visited June 6, 2018).
16   Austin, TX, Code of Ordinances, tit. 25, subchapter F, § 2.1 (2006).
17   *Id.*
18   *Id.* § 2.2.
19   *Id.* § 2.3.
20   *Id.* § 2.6.
21   *Id.*

NATIVE TREES AND INVASIVE TREES

Tyler Adams *(author)*
Jonathan Rosenbloom & Christopher Duerksen *(editors)*

INTRODUCTION

Tree mitigation ordinances seek to grow the local tree canopy by requiring the replacement of removed trees or, if that is not possible, taking alternative actions aimed at restoring green-space (see *Tree Canopy Cover*).[1] These ordinances go a step farther and require specific types of trees be used for mitigation and address the removal of smaller trees, bushes, grasses, native woodlands, and wildflowers.[2] Furthermore, ordinances can require the removal of invasive trees and their replacement with native trees.

Municipalities typically require developers or homeowners to apply for tree removal permits when seeking to remove certain types of protected trees. Protected trees can include native, heritage, and historic trees. Invasive trees are usually not considered protected, and thus may not require a permit to be removed. Further, depending on the jurisdiction, invasive trees may or may not need to be mitigated. Pursuant to this ordinance, upon approval, or as a condition of approval, protected trees removed during the course of development must be replaced from a list of approved trees. The list is frequently comprised of native trees, or at minimum, the species that have been proven to be suitable to the area. Additionally, municipalities have the option of requiring that a certain percentage of trees used as replacements be native to the area.[3] These ordinances may also identify invasive trees and require removal of those trees prior to approval.[4]

EFFECTS

Thousands of trees are destroyed every year. These trees provide critical services that are vital to many associated ecosystems and include purification of air and water, soil retention, and a variety of physical and psychological benefits. Requiring developers to replace trees removed during development with native trees, allows a community to continue to retain ecosystem benefits associated with trees. Such benefits can often be amplified with the use of native species. Trees are able to capture large amounts carbon dioxide, making them one of the most affordable and effective means of combating climate change.[5] Furthermore, they capture other pollutants, such as nitrous oxides and sulfur dioxides, which improves overall air quality.[6] Trees also aid in stormwater management and prevent further deterioration of water quality.[7] An urban tree with a 25-foot canopy with corresponding soil is capable of managing an inch of rainfall per 2,400 square feet of impervious surface through soil storage.[8] Interception and evapotranspiration also work to decrease runoff and the amount of stormwater entering municipal storm sewer systems.[9] Additionally, trees provide services in the areas of energy conservation and climate control.[10] They intercept sunlight before it is able to reach surfaces that trap heat and provide shade to buildings which cools the interior and exterior, thus reducing the amount of energy needed to artificially cool interior spaces.[11]

Requiring that replacement trees be native, or at minimum non-invasive, is especially important and can further benefit a community. Native trees are those that occur naturally in a region and are essential to protecting the biodiversity of that area.[12] Insects, birds, and other animals have co-evolved with native trees and without these native habits the associated food chain(s) is/are severed, stressing the survival of species that rely on these habits.[13] In addition, non-native trees often carry invasive pest that out compete the native species, destroying the existing habitat.[14] Because they are adapted to the soil, native trees can develop deeper and more complex root systems which help the trees hold more water, making them more efficient at stormwater management. Most native trees require less maintenance, making them less costly to maintain.[15] Typically, native trees do not require the same level of watering as non-native species and do not require fertilizer and chemicals/pesticides.[16]

EXAMPLES

Ventura County, CA

Within the coastal zone of Ventura County developers are required to protect trees and/or mitigate tree loss in order to preserve their ecological value and visual quality.[17] Keeping in place all protected trees is the most preferred approach pursuant to this ordinance, followed by on-site mitigation, off-site mitigation, and, finally, in-lieu fees.[18] When removal of a protected tree cannot be avoided, on-site mitigation requires the developer to transplant or replace the tree on the same property.[19] The standards for the replacement tree vary according to the type of tree that was removed. If one protected native tree is removed it must be replaced with no less than ten native trees.[20] Heritage trees—non-native, non-invasive, trees with unique value that are considered irreplaceable due to their rarity, distinctive feature, or prominent location—are also required to be replaced, provided they meet certain criteria.[21] If the tree is located in a public area, or prominent location as seen from public viewing areas, the replacement must be the same species as the tree removed, of similar size, and planted in close proximity to where the removal took place.[22] Heritage trees that are not viewable from or located in public areas are required to be replaced with native trees.[23]

To view the provision see Ventura County, CA, Code of Ordinances 8178-7.6 (2016).

Addison, TX

Addison, TX requires that the existing natural landscape, particularly native oak, elm, and pecan trees, be reasonable preserved.[24] The Town requires property owners to replace any dead, removed, missing, improperly pruned, or damaged trees, within thirty days of notification.[25] The list of suggested trees for replacement consist of trees native to Texas, as well as those that have been proven to be suitable to the area.[26] Trees that are not listed may be used as replacement trees subject to Addison Park Department approval.[27] Property owners may not remove or transplant a tree contained on the list without first getting a tree permit.[28] Considerations for granting removal of a listed tree include the impact removal may have on erosion, soil moisture, retention, flow of surface waters and drainage systems, and impact on the public health, safety, and welfare of the Town.[29]

To view the provision, see Addison, TX, Code of Ordinances § 34-208 (2016).

ENDNOTES

1   *See* University of Florida, *Tree Mitigation Policy*, https://perma.cc/EX8L-BGDN (last visited May 31, 2018).
2   1 Douglas W. Kmiec & Katherine Kmiec Turner, Zoning and Planning Deskbook § 5:47 (2017-2018 ed.).
3   *See* Parkland, FL, Code of Ordinances § 95-1010 (2015).
4   *See* Temple Terrace, FL, Code of Ordinances § 12-771 (2012).
5   Ecology Communications Group, *Benefits of Urban Trees*, Ecology (Oct. 31, 2012), https://perma.cc/MZQ5-NJUX.
6   *Id.*
7   Keith H. Hirokawa, *Sustainability and the Urban Forest: An Ecosystem Services Perspective*, 51 Nat. Resources J. 233, 238 (2011).
8   *Stormwater Tree: Technical Memorandum*, U.S. EPA (Sep. 2016), https://perma.cc/KE98-FKMF.
9   *Id.*
10 Hirokawa, *supra* note 7, at 239.
11 *Id.*
12 National Audubon Society, *Why Native Plants Matter*, Audubon, https://perma.cc/R5PS-EWNE (last visited June 4, 2018).
13 *Id.*
14 *Id.*
15 Scott Vogt, *Five Benefits of Native Plants*, Dyck Arboretum of the Plains (Feb. 18, 2015), https://perma.cc/AS9G-2U9K.
16 *Id.*
17 Ventura County, CA, Code of Ordinances § 8178-7.6 (2016).
18 *Id.*
19 *Id.*
20 *Id.* § 8178-7.6.1.
21 *Id.* § 8178-7.3.4
22 *Id.* § 8178-7.6.1.
23 *Id.*
24 Addison, TX, Code of Ordinances § 34-208 (2016).
25 *Id.*
26 *Id.* § 34-209; *See* Texas A&M Forest Service, *List of Trees*, https://perma.cc/C4GW-W2M9 (last visited June 13, 2018).
27 Addison, TX, Code of Ordinances § 34-209.
28 *Id.* § 34-208.
29 *Id.*

Open Space Impact Fees

Tyler Adams *(author)*
Jonathan Rosenbloom & Christopher Duerksen *(editors)*

INTRODUCTION

Impact fees are one-time charges assessed on new development in order to help pay for new or expanded public facilities and/or the impact development may have on public facilities.[1] In most jurisdictions, the fees must be used to directly address the increased demand caused by the new development and may not be used for pre-existing issues.[2] Impact fees may go toward a variety of things, such as transportation, schools, and open space. Open space is often used as an umbrella term that can incorporate parks, natural areas, conservation lands, and other outdoor recreational areas.

Local governments will typically assess and vary fees based on the type of proposed land use, such as single-family residential or multi-family residential. This ensures that uses that more densely populated or increase the burden on public facilities have a proportionally larger share of the cost associated with new development. In the ordinances, jurisdictions may also state what the proceeds will be used for, such as open space acquisition or capital investment. Even though funds must always directly address the increased demand caused by the new development, jurisdictions can place further limitations on where the funds may be spent, such as in the specific zone that the development took place. Additionally, when calculating the fees to be assessed, local governments have some flexibility, but will frequently factor in type of use, occupancy rate, number of dwellings, and/or the size of the development.[3] Some local governments will accept open space or park land dedicated to the city as credit toward the open space impact fees.

EFFECTS

Preserving open space is important for a variety of reasons and collecting impact fees allows a community to mitigate the effects of development on open space and invest in acquiring or improving existing open space. Open spaces provide essential natural habitats for certain species of wildlife and plants. Native wildlife and plants depend on undisturbed land for food, shelter, and protection, and by mitigating the loss of these lands through development impact fees, local governments can protect critical wildlife habitat and promote biodiversity.[4] Furthermore, preserving and improving open space can lead to improvements in overall air and water quality.[5] Vegetate open spaces, sometimes called green spaces, can filter stormwater by trapping and removing pollutants before they are able to enter a water resource.[6] In addition, trees help combat climate change and improve air quality by absorbing carbon dioxide and other air pollutants.[7] They also provide shade which can decrease temperatures and lead to reduced artificial cooling cost.[8]

Open spaces also allow communities to participate in a wide range of outdoor activities. This affords people the chance to interact with other members of their community and come in contact with their natural environment.[9] Studies have shown that spending time outdoors can improve physical and mental health.[10] Economically, open spaces can benefit a community as well. Studies have shown that open spaces, such as parks and recreation areas, can positively impact neighboring residential property values.[11] This can, in turn, lead to local governments receiving higher property tax revenue.[12]

EXAMPLES

Rialto, CA

Rialto established an open space development fund in order to cover the cost of acquiring new open space and any cost associated with having to construct or install infrastructure for the use of the open space.[13] Any new development that results in a decrease in the amount of public open space available in the City is assessed an open space development impact fee.[14] The funds will be used for preparing an open space master plan or other studies, including habitat conservation plan(s) "to identify and designate appropriate open space and habitat areas for preservation of threatened or endangered species."[15] Additionally, funds can be used for acquiring real property or other right of ways for the development of open space, designing and constructing improvements or facilities for the use open space, and establishing biological or cultural monitoring programs needed to maintain habitat conservation areas.[16] Rialto also has a separate parks and recreation impact fee that is imposed on all residential development projects.[17] Developers can receive credit toward the different impact fees if they construct a public facility for which a development impact fee is imposed.[18]

To view the provision, see Rialto, CA, Code of Ordinances § 3.33.230 (2013).

Deerfield Beach, FL

In order to ensure that new development carries its proportionate share of the cost of public facilities needed to accommodate it, Deerfield Beach established various development impact fees.[19] Among them is the recreation and open space impact fee. Any person seeking to get a building permit for a residential building or hotel is required to pay a recreation and open space impact fee.[20] Single-family dwellings are assessed a fee of $1,528.00, multi-family dwellings (two units or more) are assessed a fee of $1,030.00 per unit, and hotel/motel rooms are assessed a fee of $960.00 per unit.[21] This ordinance also provides exemptions to payment of impact fees when there is an alteration to an existing building that doesn't increase the residential density or adds hotel rooms and the use has not changed, or when construction of an accessory building does not result in increased demand for services by the parent parcel.[22] The funds collected from the fees are only to be used to acquire or make capital improvements to parks within the City's jurisdiction and are not permitted to be used for maintenance or operation expenses.[23] Additionally, funds not expended or encumbered seven years after the payment of the impact fee can be returned to the landowner upon request.[24]

To view the provision see Deerfield Beach, FL, Land Development Code § 98-17.1 (2016).

ADDITIONAL EXAMPLES

Bennett, CO, Municipal Code § 4-8-340 (2014) (establishing a park, recreation, and open space impact fee which is used to develop, construct, or acquire land for parks, recreation, and open space facilities).

Martin County, FL, Land Development Regulations § 6.9.J (2016) (establishing a conservation/open space impact fee to be used for conservation and open space purposes).

Durham, NC, Code of Ordinances § 30-85 (2011) (requiring impact fee funds to be spent on the capital cost of streets, parks, recreation facilities, or open space and be used exclusively in the service zone where they were collected).

Bernalillo County, NM, Code of Ordinances § 46-15 (2013) (requiring an open space impact fee for single-family and multifamily dwelling developments).

ENDNOTES

1   *Impact Fees*, MRSC Local Government Success (May 30, 2018), https://perma.cc/4ZEM-PU8Y; *Open Space Impact Fees*, Beginning with Habitat, Land Use Ordinance Tools, https://perma.cc/2CM8-YX89 (last visited July 29, 2018).

2   *Id.*

3   Because mpact fee laws and requirements vary state to state, local governments should consult state enabling legislation.

4   *Environmental Benefits*, Chicago Metropolitan Agency for Planning, Parks and Open Lands (Sep. 26, 2013), https://perma.cc/NH49-X3QL.

5   Janice C. Griffith, *Green Infrastructure: The Imperative of Open Space Preservation*, 42/43 Urb. Law. 259, 263 (Fall/Winter 2011).

6   John Heinze, *Benefits of Green Space-Recent Research*, at 10 (Environmental Health Research Foundation Aug. 25, 2011), https://perma.cc/E6GX-RJNM.

7   Griffith, *supra* note 5.

8   *See id.*

9   Andrew Chee Keng Lee et al., *Value of Urban Green Spaces in Promoting Healthy Living and Wellbeing: Prospects for Planning*, 8 Risk Mgmt. & Healthcare Pol'y 131, 132 (2015), https://perma.cc/39Y8-HXNE.

10   *Id.*

11   *Physical Activity Facilities Have Economic as Well as Health Benefits*, American Trails (Mar. 1, 2010), https://perma.cc/84K2-95WD.

12   *Id.*

13   Rialto, CA, Code of Ordinances § 3.33.230(A) (2013).

14   *Id.* § 3.33.230(B).

15   *Id.* § 3.33.230(C)(1).

16   *Id.* § 3.33.230(C).

17   *Id.* § 3.33.150.

18   *Id.* § 3.33.100.

19   Deerfield Beach, FL, Land Development Code § 98-17.1(2)(c) (2016).

20   *Id.* § 98-17.1(5)(a).

21   *Id.* § 98-17.1(5)(e).

22   *Id.* § 98-17.1(7).

23   *Id.* § 98-17.1(5)(h).

24   *Id.* § 98-17.1(6)(b).

PARKING IN-LIEU FEES

Quinn Le Frois (author)
Charlie Cowell, Tegan Jarchow, & Jonathan Rosenbloom (editors)

INTRODUCTION

Parking in-lieu fees, also known as "Park and Walk Systems," are generally used in larger cities and suburbs. They are often utilized if a defined area has sudden development growth or there are multiple establishments renovating their properties.[1] These fees are in-lieu of having or meeting individual off-street parking requirements.[2] Pursuant to ordinances with parking in-lieu fees, new developments are required to pay a fee into a designated fund.

The monies from parking in-lieu fees are used to pay solely for one or more larger parking developments[3] that serve an entire district.[4] The parking areas should be centrally located, so that pedestrians and shoppers have easy access to businesses.[5] Since the larger parking developments serve an entire district, the parking area must be strategically placed for full accessibility to all the participants within the district.[6] Depending on the jurisdiction, the parking developments are often "managed by the city, parking authority, or master developer."[7]

The in-lieu fees may be mandatory or voluntary depending on the jurisdiction.[8] Local governments can either require developers to pay the fee, or a developer can choose to pay the fee or provide individual parking.[9]

Fees are typically set using two approaches.[10] The first approach is to calculate a flat rate for each parking space not provided.[11] This flat rate is also known as a "uniform fee."[12] A uniform fee is a tangible way for developers to incorporate the proposed fee into their financial analysis.[13] The fee calculation can vary by jurisdiction, but can involve variables such as building gross square footage, gross parking demand, net parking demand, and parking provided on site.[14] The second approach is to determine appropriate development-specific fees on a case-by-case basis.[15] This approach is seen as more time-consuming and expensive because of the specific nature of each case.[16]

EFFECTS

Enacting a parking in-lieu fee ordinance can provide several benefits, including decreasing the amount of individual parking lots, which can have significant impacts on stormwater systems, energy demands, and other challenges.[17] Parking in-lieu fees provide a set option for developers. Instead of paying to construct individual parking spaces, which can be expensive, developers are given the option to pay the fee for a district parking area.[18]

In addition, because the central parking area is for the entire district, the shared parking spaces may be occupied more often when not solely devoted to one particular development or business.[19] Therefore, fewer spaces are necessary to meet the overall demand, since the turnover rate is higher when different businesses require parking at different times of the day.[20]

"[L]ess on-site parking allows continuous storefronts without gaps for adjacent surface parking lots."[21] Multiple parking lots interrupt the aesthetics of the continuous line of storefronts.[22] Parking in-lieu fees may also reduce variance requests.[23] Developers often request variances from parking minimums.[24] Because the fee includes a built-in alternative there is no need for a variance.[25] In-lieu fees also allow "adaptive reuse of historic buildings where the new use requires additional parking that is difficult to provide."[26]

Parking in-lieu fees contribute to positive pedestrian mobility systems by incentivizing people to walk. Because there are no direct off-street parking attached to each development in a parking fee district, shoppers and other pedestrians must walk from the designated parking area to their destination.[27] The presence of open shops and people on the street also encourages other people to walk.[28] Walking by shops and restaurants to get to their destination gives pedestrians a more comfortable and encouraging environment.[29] Empty streets with few pedestrians are usually avoided by other pedestrians.[30] Indeed, "[t]he more downtown is broken up and interspersed with parking lots and garages . . . 'the duller and dead it becomes'"[31]

Safety from a central parking location also benefits pedestrian mobility. With the in-lieu parking fees, there are fewer separate parking lots. Fewer parking lots can decrease the amount of traffic concentrated in a particular sector because parking is not available. Fewer parking lots also means pedestrians do not have to walk by or through parking lots and access driveways that have cars entering and exiting at different intervals (for a brief discussing the benefits of limiting access points for pedestrian safety, see Limit Driveway Access Points).[32]

Scottsdale, AZ

Scottsdale is located in Maricopa County, AZ.[33] In 2019, the population was estimated at about 246,645.[34] The City's in-lieu parking program is implemented in the downtown district.[35] The program focuses on smaller properties and provides owners opportunities to "reinvest, develop, and redevelop to the highest and best use of the property."[36] Scottsdale views the in-lieu fee policy as an investment for the future, when businesses and populations will increase. Another purpose is to provide an encouraging environment for pedestrians.[37] Fewer parking lots means fewer unsafe or inconvenient encounters that pedestrians will have with vehicles. Scottsdale also wants the ability to properly use the space in their downtown district and to fully appreciate the "urban design" of the downtown area by having fewer parking lots.[38]

Part of this in-lieu parking policy is the allocation of fees. First, a property owner or developer pays a fee to the City's downtown parking program.[39] The fees themselves are established by the City Council and subject to change.[40] Then, the fees are distributed to the downtown parking program as well as to the downtown tram service.[41] Part of the downtown parking program uses the fees for maintenance and provision of the centralized parking spots.[42]

The City Council determines whether a property owner or developer is eligible to participate in the program.[43] The City's in-lieu parking plan also includes an in-lieu parking credit system.[44] Credit is awarded by the Zoning Administrator and can be used for permanent, monthly, or evening use.[45] The number of parking credits equates to the number of parking spots available to the property owner, with the maximum being five in-lieu parking credits.[46] The property owner can "purchase and/or lease the requested number of in-lieu parking space[s]."[47] The parking

ADDITIONAL EXAMPLES

Morro Bay, CA, Code of Ordinances § 17.44.020(A)(7) (2017) (stating requirements for parking in-lieu fees and how these fees are calculated).

Dundee, FL, Code of Ordinances § 3.03.02(I)(7) (2017) (describing requirements for parking in-lieu and the limitations of the program).

Miramar, FL, Land Development Code § 715.3.1.2 (2017) (explaining in-lieu of parking fund fee calculation and the terms of the agreement for the program).

credits are used by the City Council to ensure the proper allocation of parking spots for each property owner.[48]

To view the provisions see Scottsdale, AZ, Code of Ordinances, App. B, § 9.108(D) (2018).

Dania Beach, FL

Dania Beach is located in Broward County, FL.[49] In 2017, the estimated population was 32,030.[50] The Community Development Director determines the adequacy of the current on-site parking situations at new or existing developments.[51] If there is inadequate on-site parking, the owner of the property applies for a waiver allowing off-street parking spaces.[52] If the waiver is granted, the property owner pays a fee-in-lieu of the on-site parking.[53] The fees are calculated by the "average cost to the city for the construction of a parking space in a parking structure."[54] Collectively, these decisions are made by the Director of Finance, the Public Services Director and the Community Development Director.[55] The City also has Parking Priority Districts; these districts receive priority use of the parking in-lieu funds.[56]

To view the provisions see Dania Beach, FL, Code of Ordinance § 265-92 (2012).

ENDNOTES

1   Robin Zimbler, Driving Urban Environments: Smart Growth Parking Best Practices 4, https://perma.cc/3T7E-AXAJ (last visited May 28, 2019).
2   Jeff Speck, Walkable City Rules – 101 Steps To Making Better Places 40 (2018).
3   Naperville, IL, Code of Ordinances § 11-2E-3 (2018), https://perma.cc/787Z-HA5W.
4   Speck, *supra* note 2, at 40.
5   *Id.*
6   *Id.*
7   *Id.*
8   Kimley-Horn & Assocs., Downtown Boise Parking Strategic Plan: Appendix 12 – Parking Planning White Paper Series, Parking in Lieu Fees 9 (Capital City Development Corp. Dec. 2012), https://perma.cc/T6P6-QEK6.
9   Donald C. Shoup, In Lieu of Required Parking 309 (2001), https://escholarship.org/uc/item/8rp7s63c.
10   Zimbler, *supra* note 1, at 4.
11   *Id.*
12   Shoup, *supra* note 9, at 309.
13   Columbus Short North Parking Study, Best Practices-Parking in Lieu Fee 14, https://perma.cc/4JEU-GZEQ (last visited May 28, 2019).
14   *See e.g.*, Naperville, IL, Code of Ordinances § 11-2E-3.
15   Zimbler, *supra* note 1, at 4.

16   Columbus Short North Parking Study, *supra* note 13, at 14.

17   *Id.*

18   Shoup, *supra* note 9, at 308.

19   *Id.*

20   *Id.*

21   *Id.*

22   *See id.*

23   *Id.*

24   *Id.*

25   *Id.*

26   *Id.*

27   Michael Manville & Donald Shoup, *People, Parking, and Cities*, Access (Fall 2004), https://perma.cc/FRW5-WX7P.

28   *Id.*

29   *Id.*

30   *Id.*

31   *Id.*

32   *See* Zimbler, *supra* note 1, at 25.

33   World Pop. Rev., *Scottsdale, Arizona Population 2019* (last updated May 27, 2019), https://perma.cc/7SNT-ET9D.

34   *Id.*

35   *See* Scottsdale, AZ, Code of Ordinances, App. B, § 9.108(D) (2018).

36   *Id.*

37   *Id.*

38   *Id.*

39   *Id.*

40   *Id.*

41   *Id.*

42   *Id.*

43   *Id.*

44   *Id.*

45   *Id.*

46   *Id.*

47   City of Scottsdale, In-Lieu Parking: Development Application Checklist 2 (Mar. 2, 2015), https://perma.cc/PY3J-CCPT.

48   *Id.*

49   Dania Beach, FL, Code of Ordinance § 265-92 (2012).

50   *Id.*

51   *Id.*

52   *Id.*

53   *Id.*

54   *Id.*

55   *Id.*

56   *Id.*

Parking Maximums

Brandon Hanson (author)
Jonathan Rosenbloom & Christopher Duerksen (editors)

INTRODUCTION

Off-street parking maximum standards in zoning ordinances limit the construction of parking lots that are larger than necessary. Local governments across the U.S. have routinely set parking *minimums* in their land development regulations for various types of uses.[1] The purpose of parking minimums is to insure that there are sufficient off-street parking spaces for each development based, typically, on the building use and size.[2] Increasingly, local governments recognize the need to limit parking for a variety of reasons and therefore establish parking *maximums* in their regulations, establishing an upper bound for the number of spaces allowed for a specific use, thus controlling the amount of land and impervious surface associated with parking.[3] Some jurisdictions permit developments to exceed maximums upon the performance of certain criteria, such as increasing permeable surfaces.

EFFECTS

While off-street parking is an aspect of most developments, impervious parking lots can have a number of detrimental effects. First, they prevent groundwater infiltration and increase storm water run-off.[4] Such run-off can increase downstream erosion and flooding as well as polluting rivers and lakes. Additionally, impervious parking lots may impose significant costs and burdens on municipal stormwater systems. They may also increase greenhouse gas (GHG) emissions where stormwater must be treated by a local utilities to remove pollutants to meet federal and state standards.[5] Second, large parking lots can reduce densities and spread out development over a larger area, thereby hindering the use of less GHG-intensive modes of transportation, such as walking, biking, or public transit.[6] Third, cheap and convenient off-street parking promotes increased use of vehicles[7] that can lead to traffic congestion, air pollution, and poor public health.[8] Traffic congestion in turn results in calls for wider streets, bigger intersections, and, ironically, even higher parking requirements.[9] Finally, large parking lots with their expansive pavement can exacerbate the urban heat island effect, making local communities hotter and increasing demand on energy.

Importantly, parking maximums may also result in less excessively large and wasteful parking lots. When those spots are under-utilized, consumers, developers, and local governments are paying unnecessary charges.[10] Parking maximum standards can reduce the physical size of lots thereby promoting compact developments while reducing stormwater run-off and GHG emissions.[11] While minimum parking requirements have produced no single disaster, "evidence of their harm confronts us everywhere-traffic congestion, air pollution, energy imports, the orientation of the built environment around the automobile, perhaps even global climate change. Although not their sole cause, minimum parking requirements magnify all these problems."[12]

EXAMPLES

Portland, OR

Portland enacted a parking maximum ordinance by creating multiple formulas for different use categories.[13] Each use category is set out in a table accompanied by a ratio used to determine the parking minimum as well as the parking maximum.[14] Commercial spaces are usually subject to a cap based on their floor area.[15] For example, a general office category is allowed one parking space per 294 square feet of office space, maximum.[16][16] Thus, using the ratio described above, a project with a proposed office space of 5,000 square feet would result in ten parking slots

ADDITIONAL EXAMPLES

Charlotte, NC, Code of Ordinances, Zoning, § 9.1208 (6) (2018) (setting parking minimums and maximums in transit-oriented districts only).

Flagstaff, AZ, Zoning Code § 10-50.80.040 (C) (1) (2018) (setting a maximum amount of parking at five percent higher than the minimum).

Vancouver, Canada, Parking Bylaws § 4 (2019) (implementing conventional parking maximums as wells as a total parking cap in the downtown area).

Denver, CO, Municipal Code § 30-50 (2018) (requiring developer to ask special permission to include parking above the parking minimum).

New Haven, CT, Zoning Ordinances § 45 (8) (D) (2018) (providing for a parking maximum in mixed-use districts of three spaces per 1,000 square feet).

as a maximum (5,000 / 294 = 17.01). In some cases, Portland reduces its minimum and maximum lot sizes if the development is located close to a transit system.[17]

To view the provision see Portland, OR, City Code § 33.266.115 (2018).

Hartford, CT

Hartford manages parking lot sizes by setting out parking maximums though a table of uses classifications.[18] The code sets out specific maximum numbers of parking spaces for all of its allowed use classifications.[19] Developers must determine which use category a development falls into. For example, a restaurant would qualify as a place of eating.[20] This category allows a maximum of three parking spaces for every five people of the restaurant's maximum capacity.[21] If the restaurant had a maximum of 100 customers, then the maximum parking spaces would be 60 parking spaces. This helps Hartford control the amount traffic and vehicle use in and out of the developments in question. Hartford has become one of the first cities in the country to eliminate minimum vehicular parking requirements. Instead, Hartford imposes minimum bicycle parking requirements. Together with the parking maximums, Hartford is repositioning itself for a future with less dependence on the automobile.

To view the provision see Hartford, CT, Zoning Regulations § 7.2.2 (B) (2018).

> ### ADDITIONAL EXAMPLES (cont'd.)
>
> Burlington, MA, Zoning Bylaw §§ 7.2.4, 7.2.5 (2015) (setting both parking minimums and maximums for various types of developments).
>
> Knoxville, TN, Zoning Regulations Art. 5 § 7 (D) (2018) (creating parking minimums and maximums with exceptions from the department of engineering).
>
> New York, NY, Zoning Resolution Art. 2 Ch. 5 (2018) (creating parking maximums, no minimum requirements for specific buildings).

ENDNOTES

1   *See, e.g.,* Flagstaff, AZ, Zoning Code § 10-50.80.040; Omaha, NE, Code of Ordinances § 55-734; Yakima, WA, Code of Ordinances Table 6-1; Coppell, TX, Code of Ordinances § 12-31-6; RICHARD W. WILLSON, PARKING REFORM MADE EASY 11-12 (2013).

2   WILLSON, *supra* note 1, at 20-21 (2013); CHRISTOPHER MCCAHILL & NORMAN GARRICK, PARKING: ISSUES AND POLICIES, PARKING SUPPLY AND URBAN IMPACTS 35-37 (Stephen Ison & Corinne Mulley ed., 2014).

3   WILLSON, *supra* note 1, at 34-36; Portland, OR, City Code § 33.266.115; Harford, CT, Zoning Regulations Fig. 7.2-A; Burlington, MA, Zoning Bylaws §§ 7.2.4, 7.2.5.

4   ANDREW KAVARVONEN, POLITICS OF URBAN RUNOFF: NATURE, TECHNOLOGY AND THE SUSTAINABLE CITY 10-11 (2011); Benjamin O. Brattebo & Derek B. Booth, *Long-term Stormwater Quantity and Quality Performance of Permeable Pavement Systems*, 37 WATER RESEARCH 4369, 4369 (2003).

5   KAVARVONEN, *supra* note 4, at 13-14.

6   McCAHILL & GARRICK, *supra* note 2, at 40-41.

7   GREG MARSDEN, PARKING: ISSUES AND POLICES, PARKING POLICY 17-18 (Stephen Ison & Corinne Mulley ed., 2014).

8   *Id.* at 17-18; WILLSON, *supra* note 1, at 24-29; Michael Lewyn, *Sprawl in Canada and the United States*, 44 URBAN LAWYER 85 (Winter 2012).

9   Donald C. Shoup, *The Trouble with Minimum Parking Requirements*, 33 TRANSP. RES. PART A 549 (1999).

10 *Id.* at 1.

11 MARSDEN, *supra* note 7, at 17-18.

12 Shoup, *supra* note 9.

13 Portland, OR, City Code § 33.266.115.

14 *Id.* at tbl. 266-2

15 *Id.*

16 *Id.*

17 *Id.* § 33.266.115 (B) (1) (b).

18 Harford, CT, Zoning Regulations Fig. 7.2-A.

19 *Id.*

20 *Id.* at fig. 7.2-A.

21 *Id.*

SAFE ROUTES

Nicole Steddom (author)

Jonathan Rosenbloom & Christopher Duerksen (editors)

INTRODUCTION

Creating protected walking paths to school, work, and nature is vital to make walking and biking more attractive. When people walk more, they are healthier, and they reduce greenhouse gas (GHG) emissions and other pollutants associated with driving gas-powered cars. This ordinance seeks to increase walking by requiring or encouraging the development of "Safe Routes." Safe Routes is a national initiative focused on creating safe pathways for children to get to school by walking or biking.[1] Building off the national Safe Routes campaign, many municipalities have embedded safe routes into their codes and creatively expanded walking and biking routes in order to benefit a broader array of citizens.

Local governments may design these ordinances to require the creation of safe routes or they may create incentives to encourage developers and home owners to incorporate safe routes into the properties. Many communities have realized the environmental benefit of reducing the number of vehicles on the road, and have simply required that developers provide trails or other walkways to local schools.[2] Other municipalities require developers to perform a safe routes analysis, while others require schools to be built in particular places or set maximum walking distances from neighborhoods to schools.[3] Some local governments have incentivized the creation of safe routes by making them a requirement in order to receive certain incentives, certificates of occupancy, or building approvals.[4]

EFFECTS

Creating safe routes to school, work, and nature encourages walking and bicycling which not only improves citizens' health, but it also serves to reduce GHG emissions by giving safe alternatives to vehicular transport. Safety is of course a first priority. Pedestrians have a two times greater likelihood of being hit by a car where there are no sidewalks.[5] In 2009, approximately 23,000 children were injured or killed while walking or bicycling in the U.S.[6] Medical costs for children and their families were around $839 million in 2005 when treating children's bicycling and walking fatalities.[7]

The second priority for creating safe routes to work, school, and nature is health. Two miles of walking is the equivalent of two-thirds of the recommended daily activity for children.[8] Obesity is on the rise for children and adults. Creating an incentive to do something as simple as bicycling or walking can make a significant difference in rates of obesity. It is estimated that one-fourth of health care costs in the U.S. can be attributed to obesity.[9] The cost can be almost $14 billion a year for childhood obesity.[10] Walking and/or bicycling can also improve a child's academic performance. Studies have shown that children who engage in physical activity before class are more attentive, learn better, and experience an increase in memory.[11]

The third priority for creating safe routes to work, school, and nature is environmental. According to Safe Routes Partnership, "returning to 1969 levels of walking and bicycling to school would save 3.2 billion vehicle miles, 1.5 million tons of carbon dioxide and 89,000 tons of other pollutants—equal to keeping more than 250,000 cars off the road for a year."[12] Although development and sprawl over the last fifty years may make a return to 1969 levels extremely difficult, even a 5% increase in "walkability" for a neighborhood can reduce vehicular miles by 6%.[13] It is estimated that one-third of schools are in "air pollution danger zones."[14] Asthma is on the rise in children, and children that are routinely exposed to traffic pollution have a much greater likelihood of developing heart and lung problems as adults.[15] Creating safe routes for children to walk to school and nature can reverse these trends, while saving money.

EXAMPLES

Cibolo, TX

In Cibolo's Code of Ordinances, there are a wide variety of sustainable goals.[16] The City's application of mixed-use development in this particular ordinance focuses on transportation networks and how pedestrians and bike enthusiasts can coexist in a manner that is efficient for vehicles, and attractive for pedestrians or bicyclists.[17] Pertinent to the idea of Cibolo's safe routes is the concept of block design.[18] The overall goal of block design is primarily to "create pedestrian-oriented development by establishing a well-defined pattern of walkable blocks and intersecting streets, attractive and well-designed streetscapes that are human-scaled and pedestrian friendly."[19]

There is also an extensive discussion in the code that deals strictly with pedestrian and bicycle access.[20] Cibolo has created standards that must be complied with in all new development such as access from the site of development to public bike paths or greenways.[21] Other standards require sidewalks through parking lots, primary entrances to buildings with connections to greenways or trail systems, and connections between developments, adjacent uses, and perimeter sidewalks. These pathways must have direct pedestrian and bicycle compatibility.[22]

To view the provision see Cibolo, TX, Code of Ordinances § 4.7 (2017).

Pima County, AZ

Like many jurisdictions, Pima County, Arizona provides general standards for subdivisions.[23] In Pima County, however, there must be a safe route linkage between schools and subdivisions. Those linkages must be one and a half miles for elementary schools and two and a half miles for middle schools.[24] Pima County requires that developers submit safe routes plans to school boards. Such plans must be submitted after having met with the school district and discussed the plat or development plan.[25] There must be ten acres set aside in the subdivision for public recreation areas. Further, design and development must be sensitive to the native environment and may include trails.[26] Recreation area plans must be submitted for review at thirty, sixty, ninety, and one hundred percent completion stages.[27]

To view the provision see Pima County, AZ, Code of Ordinances § 18.69.090 (2017).

ADDITIONAL RESOURCES

UNC Highway Safety Research Center, *Safe Routes: National Center for Safe Routes to School*, https://perma.cc/BY3G-CCDM (last visited July 2, 2018).

Safe Routes to School National Partnership, *Healthy Communities: Quick Facts and Stats*, https://perma.cc/JYU8-39QV (last visited July 12, 2018) (information about Safe Routes for school).

National Recreation and Parks Association, *Safe Routes to Parks*, https://perma.cc/DGN8-U5BN (last visited July 12, 2018) (initiative to create routes to parks).

Collier County Government, *Rural Land Stewardship Area (RLSA) Overlay Program*, https://perma.cc/3CQ8-XSCA (last visited July 2, 2018) (explaining the RLSA program in Collier County).

U.S. DEPARTMENT OF TRANSPORTATION, A RESIDENT'S GUIDE FOR CREATING SAFER COMMUNITIES FOR WALKING AND BIKING (Jan. 2015), https://perma.cc/WMT4-N2F7 (describing steps that can be taken by residents to increase walking and biking in their community).

Local and Regional Government Alliance on Race & Equity, *Racial Equity Toolkit: An Opportunity to Operationalize Equity* (Oct. 30, 2015), https://perma.cc/CDN7-5Q7Y (outlining ways in which local governments can create policies to eliminate racial inequality).

ChangeLab Solutions, *Complete Parks: Creating and Equitable Parks System*, https://perma.cc/8V6M-CCN3 (last visited July 2, 2018) (tools to create equitable park systems in all neighborhoods).

ChangeLab Solutions, *A Guide to Building Healthy Streets: How Public Health Can Implement Complete Streets*, https://perma.cc/L33K-VUGM (last visited July 2, 2018) (providing tools to make streets safer and more accessible for bikers and walkers).

Safe Routes to School National Partnership, *Safe Routes to School Meets Safe Routes to Parks* (Dec. 17, 2015), https://perma.cc/P943-7W2L (strategies to increase physical activity in the lives of residents by increasing the safety of streets for bikes and pedestrians).

ENDNOTES

1 UNC Highway Safety Research Center, *Safe Routes: National Center for Safe Routes to School*, https://perma.cc/BY3G-CCDM (last visited July 2, 2018).
2 Pima County, AZ, Code of Ordinances § 18.69.090 (2017).
3 McKinney, TX, Code of Ordinances § B-2 (2018).
4 St. Lucie County, FL, Land Development Code § 4.05.08 (2017).
5 Safe Routes to School National Partnership, *Healthy Communities: Quick Facts and Stats*, https://perma.cc/JYU8-39QV.
6 *Id.*
7 *Id.*
8 *Id.*
9 *Id.*
10 *Id.*
11 *Id.*
12 *Id.*
13 *Id.*
14 *Id.*
15 *Id.*
16 Cibolo, TX, Code of Ordinances § 4.7 (2017).
17 *Id.*
18 *Id.*
19 *Id.*
20 *Id.*
21 *Id.*
22 *Id.*
23 Pima County, AZ, Code of Ordinances § 18.69.090 (2017).
24 *Id.*
25 *Id.*
26 *Id.*
27 *Id.*

SITE & SOLAR ORIENTATION

Glenn Holmes, Bradley Adams (authors)
Darcie White, Sara Bronin, & Jonathan Rosenbloom (editors)

INTRODUCTION

The design, orientation, and layout of a structure directly affect the efficiency of solar energy generation.[1] Solar energy regulations may require solar-ready lot and building orientation,[2] and site plan regulations may require a site layout that provides a minimum length of time for solar energy systems to have access to sunlight.[3] Optimum solar capacity can be achieved, in part, by adopting regulations that require street and building orientation to have the long axis running from east to west in order to maximize sunlight exposure emanating from the south.[4] Codes can allow flexibility in their setback requirements, as well as limit the height and location of structures, so as not to interfere with a development's capability to harvest and produce solar power.[5] Communities that adopt solar siting ordinances into their site development standards may control the height and location of buildings through stricter minimum setbacks to ensure neighboring lots maintain solar access.[6]

The effectiveness of site orientation is heavily dependent on ordinances, which combine the concept with shade prevention or "solar fences."[7] For ordinances addressing off-property shading and off-property shading disputes see Process to Resolve Tree Interference with Solar Access brief and Limiting Off Property Shading of Solar Energy Systems brief.

EFFECTS

Site orientation is a critical component to "passive" solar energy collection, a process that involves harnessing solar energy to heat a building without the use of panels or other instruments. The ideal orientation for buildings to harvest solar energy is within five degrees of true south, though placement within 30 degrees still garners a considerable solar contribution.[8] Aligning a house in accordance with the recommended axis, combined with design features such as thermal mass material and the installation of glazed southern facing windows, can reduce heating costs for homeowners by as much as 25 percent without the need to install additional solar energy collection equipment.[9] Proper placement of sunspaces or glass rooms on the southern end of properly aligned structures can account for up to 60 percent of a residence's heating.[10] Placing an awning or similar covering instrument over windows in homes utilizing passive solar energy reduces cooling costs in the summer because of the sun's higher trajectory.[11]

Using photovoltaics (PV) and other solar collectors to store energy for later use is known as "active" solar energy collection. The orientation of a structure affects the effectiveness of solar energy equipment.[12] Solar panels on roofs facing the east or west will return approximately 20 percent less energy than south facing panels.[13] North facing panels offer little to no energy production.[14] An array aligned to face south reduces the amount of time it takes for a PV system to provide a full return on the cost of the investment.[15] In one study, an east facing, 10 kilowatt PV array in Golden, Colorado increased the simple payback period by over three years over a similar north facing array, while a west facing array increased the payback period by over five years.[16]

Building a home with the long walls aligned on the east-west axis provides the residents with a better perception of cardinal direction, time, weather conditions, and aids understanding of "the world outside the building."[17] The costs associated with constructing buildings with elongated south facing walls are minimal, and can be outweighed by the positive effects of lighting and energy consumption.[18] Utilization of passive solar carries the additional benefit of promoting health and adding comfort to buildings by reducing dependency on furnaces. Forced air heating systems used in typical construction create discomfort for occupants by pulling dry air from outside of the structure, significantly decreasing the level of humidity inside.[19] This can lead to a sensation of dry skin, while also creating an environment conducive to the well-being of viruses and bacteria.[20] By requiring only sunlight and proper design, passive solar heating can circumvent the risks associated with traditional heating systems.

EXAMPLES

Boulder, CO

Boulder is divided into three "Solar Access Areas."[21] Areas I and II protect existing access to sunlight for south facing yards, walls and rooftops, while Area III calls for a permit process to be employed when granting unfettered solar access would unduly hinder development.[22] The City utilizes solar fences to preserve the right to sunlight in Solar Access Areas I and II.[23] Flexibility is offered in the form of an application for an exception when developers wish to deviate from the code's solar requirements.[24]

Solar siting requirements for new subdivisions and planned unit developments require residential units to have a roof surface oriented within thirty degrees of a true east-west direction and to be either flat or sloped in a southern direction.[25] Each residential unit in a Solar Access Area must also have an exterior wall surface that is oriented within thirty degrees of a true east-west direction and located on the southernmost side of the unit.[26] All nonresidential buildings with an anticipated hot water demand of one thousand or more gallons per day must have a roof surface that is flat or oriented within thirty degrees of a true east-west direction.[27]

> **ADDITIONAL EXAMPLES**
>
> North Las Vegas, NV, Code of Ordinances § 17.24.140-1: Menu of Sustainability Options (offering density bonuses through a point system, including for orienting lots and streets to maximize solar efficiency).
>
> Anchorage, AK, Code of Ordinances § 21.07.110(C)(8)(e) (requiring residences with eight or more units to select three items from a list, including the provision of windows or primary entrances on twenty percent of a solar oriented wall which is likely to have six or more hours of sunlight from March through September).

When subdividing land, Boulder's municipal code requires subdivision plats to maximize solar energy use by requiring solar siting criteria for new subdivisions.[28] The development must orient streets, lots, open spaces and buildings to ensure that each principal building on the lot is able to realize its solar potential.[29] The code also requires lots to be designed to avoid shading by nearby structures and minimize off site shading of adjacent properties.[30] Additionally, developers must design the building shape to maximize utilization of solar energy and must utilize open space areas to protect buildings from shading by other structures.[31]

To view the provisions see Boulder, CO, Municipal Code §§ 9-9-17 and 9-12-12(a)(1)(O)(i-iv).

Eatonton, GA

Eatonton allows rooftop solar energy collectors as a by-right accessory structure in every district with the exception of its Downtown Business Overlay zone.[32] In the Downtown Business Overlay zone, solar energy collectors are permitted pursuant to the regulations of the underlying district.[33] The City requires that new structures be built ready to exploit solar energy, and developers must take into account building orientation, the effects of shading and wind-screening by vegetation, and the possibility of shading nearby properties.[34] Streets in a new development are to be aligned on an east west axis where feasible to encourage maximum sunlight exposure when erecting new buildings.[35]

ADDITIONAL RESOURCES

AMERICAN PLANNING ASSOCIATION, PLANNING FOR SOLAR ENERGY BRIEFING PAPERS (2013), https://perma.cc/VE54-W5PE (providing a comprehensive overview of many solar energy related topics).

John R. Nolon, *Mitigating Climate Change by Zoning for Solar Energy Systems: Embracing Clean Energy Technology in Zoning's Centennial Year*, Pace Law Faculty Publications (Dec. 2015), https://perma.cc/5ESR-DAWT (detailing strategies for local governments to promote solar energy).

Within reasonable bounds, new subdivisions should be platted to maintain or improve the quality of active and passive solar energy systems.[36] Cluster subdivisions must be designed to promote solar energy for dwellings, and should consider placing high-density units on south facing slopes and low-density units on north facing slopes.[37] To reduce shading, structures should be placed on the northern end of lots, and larger structures should be placed on the northern side of smaller ones.[38] Eatonton's planning and zoning commission is required to consider if approval of new projects will block an existing solar collector's access to sunlight during peak daylight hours.[39] Both the City and the planning and zoning commission have the right to reject platting and subdivision plans that are prohibitive to the reasonable development of solar collectors or other renewable energy devices.[40]

To view the provision see Eatonton, GA, Code of Ordinances §§ 53-4, 53-(9-10).

ENDNOTES

1   John R. Nolon, *Mitigating Climate Change by Zoning for Solar Energy Systems: Embracing Clean Energy Technology in Zoning's Centennial Year*, Pace Law Faculty Publications 6-7, 27 (Dec. 2015), https://perma.cc/5ESR-DAWT.

2   *Id.*

3   AMERICAN PLANNING ASSOCIATION, PLANNING FOR SOLAR ENERGY BRIEFING PAPERS 41 (2013), https://perma.cc/VE54-W5PE.

4   *Id.* at 36

5   *Id.*

6   *Id.*
7   *See, e.g.* Boulder, CO, Municipal Code § 9-9-17(d)(1)(A-B) (2013).
8   Passive Solar Industries Council, *Passive Solar Design Strategies: Guidelines for Home Building, National Renewable Energy Laboratory* 16, https://perma.cc/M55E-BG54 (last visited Jun. 18, 2019).
9   *What's in an Environmentally Responsible Building?*, Union of Concerned Scientists, https://perma.cc/72TS-H9S8 (last visited Jun. 18, 2019).
10   *Solar Energy*, Environmental and Energy Study Institute, https://perma.cc/4Z6J-AM4B (last visited Jun. 18, 2019).
11   *What's in an Environmentally Responsible Building?, supra* note 9.
12   Patrina Eiffert & Gregory J. Kiss, Building-Integrated Photovoltaic Designs for Commercial and Institutional Structures 60 (National Renewable Energy Laboratory 2000), https://perma.cc/LX8X-M82W.
13   *Impact of roof orientation on solar savings*, Energysage (2019), https://perma.cc/35RT-ATLE.
14   *Id.*
15   Andrea Watson et al., Solar Ready: An Overview of Implementation Practices, National Renewable Energy Laboratory 4 (2012), https://perma.cc/CSQ7-XNBB.
16   *Id.*
17   Shahrzad Fadaei et al., The Effects of Orientation and Elongation on the Price of the Homes in Central Pennsylvania 2 (Penn. State University Dep't of Architecture 2015), https://perma.cc/XAN7-8PSS.
18   *See id.* at 8.
19   *Humidity*, Home Energy Center (2018), https://perma.cc/899Q-EJEW.
20   *Id.*
21   Boulder, CO, Municipal Code § 9-9-17(c)(1-3).
22   *Id.*
23   *Id.* § 9-9-17(d)(1)(A-B).
24   *Id.* § 9-9-17(f).
25   *Id.* § 9-9-17(g)(1)(A)(i-ii).
26   *Id.* § 9-9-17(g)(1)(B)(i-ii).
27   *Id.* § 9-9-17(g)(1)(C)(i).
28   *Id.* § 9-12-12(a)(1)(O)(i-iv).
29   *Id.*
30   *Id.*
31   *Id.*
32   Eatonton, GA, Code of Ordinances § 53-4 (2016).
33   *Id.*
34   *Id.* § 53-9(a).
35   *Id.* § 53-9(b).
36   *Id.* § 53-9(e).
37   *Id.* § 53-9 (f)(1).
38   *Id.* § 53-9(f)(2-3).
39   *Id.* § 53-9(d).
40   *Id.* § 53-10.

SOLAR-READY CONSTRUCTION

Bradley Adams (author)

Darcie White, Sara Bronin, & Jonathan Rosenbloom (editors)

INTRODUCTION

This ordinance requires developers to construct solar-ready homes and commercial buildings. Homeowners and commercial property owners who want to install solar energy collectors on existing structures may encounter cost prohibitive hurdles due to incompatible building and roof design, inadequate electrical components, reduced capacity resulting from shading and orientation, and high cost of retrofitting solar equipment on an existing building.[1] When drafting solar ready construction ordinances local governments should consider a variety of issues, including site orientation, roof design, shading, and mechanical specifications, such as electrical service and plumbing.[2]

Site orientation and shading analysis are critical components of "passive" solar energy collection. Passive structures are designed to maximize sunlight for heating without the use of external equipment such as photovoltaic (PV) or solar water heater (SWH) instruments. "Active" solar energy collection utilizes PV and SWH apparatuses to convert sunlight into electricity or fuel for hot water or other purposes (for a briefs on protecting solar energy systems from shading and for disputes when such shading occurs see Limiting Off Property Shading of Solar Energy Systems and Process to Resolve Tree Interference with Solar Access). PV and SWH systems benefit from south-facing site orientation and protection from shading, but in order to maximize a building's orientation and make it more solar-ready, additional steps are necessary in the construction phase (for a description of the importance of site orientation, see Site & Solar Orientation brief).

To be solar-ready, rooftops should be properly angled for the specific location and/or sloped toward the south, and should be able to handle the equipment weight and weather conditions after installation.[3] All collateral materials on the roof, such as vents, chimneys, and mechanical equipment, should be grouped to reserve as much space as possible for solar collectors.[4] For larger SWH systems, consideration should be given to ensuring that the proper plumbing system is installed, connecting the rooftop to the equipment room at the time of construction.[5] Finally, solar-ready PV energy collectors require forethought on how the system will connect to the grid, the location of the electric panel on the structure, as well as the size and output of the array.[6]

This ordinance may work well with other solar-ready ordinance briefs, such as Allow Solar Energy Systems and Wind Turbines by-Right, Change Height & Setbacks to Encourage Renewables, and Streamline Solar Permitting and Inspection Processes.

EFFECTS

Incorporating solar-ready specifications into building designs creates substantial savings for owners on installation costs and may encourage such installation.[7] Installation costs vary geographically, but research estimates that the cost of a 10 kilowatt PV energy collector can be cut by $2,644 if a structure is built to accommodate photovoltaic systems.[8] Retrofitting a building to install an SWH system impacts cost through the placement of copper pipes running from the rooftop to the equipment room, the installation of tee joints to connect the system to the plumbing, and mounting the SWH hardware to the roof.[9] As is the case with PV systems, costs vary by specific site, but research shows that for a three-panel, 96 square foot SWH system, owners can expect to save $3,057 on installation costs when designing a building to be SWH ready.[10] Savings increase as building and system sizes are scaled up. For example, up front installation costs for PV on a three-story mixed-use building can range from $5,000-$7,500, while retrofitting that same building to accommodate solar energy ranges from $20,000-$30,000.[11]

Roof design and orientation substantially impact the amount of energy that solar collectors are able to harvest. Arrays that are aligned to the east and west are 20 percent less effective than those facing south.[12] Panels atop roofs that are north-facing produce little energy, and are likely not worthwhile.[13] Rooftop orientation is a low-cost way to support solar implementation and create positive investment returns when solar energy systems are installed.[14]

Pinecrest, FL

Pinecrest requires developers of new construction or remodeling that exceeds 50 percent of the appraised building value to set forth a design indicating how the roof will be built to accommodate a PV or SWH system.[15] The ultimate owner is responsible for the eventual installation of either a PV or SWH system.[16] Pinecrest's code requires new homes that exceed 6,000 square feet to also include several conservation-based features.[17] Requirements include the installation of either a solar or tankless water heater, a hybrid electric water heater, or a PV system.[18] The code mandates that structures have an auxiliary energy system, which can be accomplished through the installation of a solar panel and battery pack.[19] Before a certificate of occupancy is issued, the general contractor must provide an affidavit stating that the building complies with the energy code.[20]

To view the provision see Pinecrest, FL, Code of Ordinances Ch. 30 Art. 5 Div. 5.27(a)(3)(a-c).

El Paso, TX

El Paso adopted both the solar-ready provisions of Appendix U from the 2015 International Residential Code (IRC) for townhouses and detached one and two family dwellings and the Recommended Requirements to Code Officials for Solar Heating, Cooling and Hot Water Systems.[21] The IRC provision states that new townhomes and detached one or two-family homes aligned between 110 and 270 degrees of true north, which contain 600 or more square feet of roof space, comply with the solar-ready requirements.[22] The code mandates that the area of rooftops designated for solar access be free from obstructions such as chimneys and vents.[23] Construction documents must set

ADDITIONAL EXAMPLES

South Salt Lake, UT, Code of Ordinances § 15.12.840(G) (2011) (requiring new developments to be solar ready).

Long Beach, CA, Municipal Code § 21.45.400(I)(3) (2019) (requiring roofs be solar ready by being able to bear an additional eight pounds of dead load per square foot. A conduit must also be provided from the electrical panel to the roof).

Louisville, CO, Code of Ordinances § 15.05.020 (2018) (adopting the 2018 version of the International Residential Code's solar ready provisions)..

forth the design for electrical and plumbing components to connect to PV or SWH systems,[24] and space must be set aside on the main electrical service panel to allow for the future installation of solar equipment.[25]

Pursuant to the Recommended Requirements to Code Officials for Solar Heating, Cooling and Hot Water Systems, buildings and equipment thereon must be able to support the weight of the energy collecting system and all other static ("dead load") and fungible ("live load") elements on the structure.[26] In addition, the equipment must be able to endure natural elements such as snow, wind, expansion and contraction due to temperature change, and seismic activity.[27]

To view the provisions see City of El Paso, TX, Code of Ordinances §§ 18.10.366, 18.32.010 (2016); 2015 Int'l Residential Code, Int'l Code Council App. U § U103.1 (2016); Recommended Requirements to Code Officials for Solar Heating, Cooling and Hot Water Systems, Council of American Building Officials 19 (1980).

ADDITIONAL RESOURCES

Lisell, T. Tetreault, & A. Watson, Solar Ready Buildings Planning Guide, National Renewable Energy Laboratory (2009), https://perma.cc/6T49-SEYF (offering a comprehensive checklist for solar-ready building design).

Lunning Wende Associates, Inc., Solar Ready Building Design Guidelines, The Minneapolis Saint Paul Solar Cities Program (Sept. 2010), https://perma.cc/U24H-93QM (providing easily accessible guidelines for solar-ready buildings).

ENDNOTES

1 ANDREA WATSON ET AL., SOLAR READY: AN OVERVIEW OF IMPLEMENTATION PRACTICES, NATIONAL RENEWABLE ENERGY LABORATORY 3-5, 9-10 (2012), https://perma.cc/68ZQ-NHYM. The National Renewable Energy Lab has created a comprehensive checklist for designing solar-ready buildings (to view the full checklist, visit the NREL website listed below in Additional Resources). L. LISELL, T. TETREAULT, & A. WATSON, SOLAR READY BUILDINGS PLANNING GUIDE 3-6 (National Renewable Energy Laboratory 2009), https://perma.cc/A8QW-QVKX.
2 Id. at 4-6.
3 Id. at 3.
4 Id.
5 Id. at 21.
6 Id. at 24-28.
7 SOLAR READY KC SOLAR INSTALLATION POLICY AND PRACTICE IN KANSAS CITY AND BEYOND, MID-AMERICA REGIONAL COUNCIL 21 (2013), https://perma.cc/4478-6X27.
8 WATSON ET AL., supra note 1, at 5.
9 Id. at 7.
10 Id.
11 N'ATL ASS'N OF REG'L COUNCILS ,BEST MGMT. PRACTICES FOR SOLAR INSTALLATION POLICY 3, https://perma.cc/B4HA-KBDE (last visited June 17, 2019).

12   Impact of roof orientation on solar savings, Energysage (2019), https://perma.cc/X4JZ-4Q3P.
13   *See id.*
14   *See* SHAHRZAD FADAEI ET AL., THE EFFECTS OF ORIENTATION AND ELONGATION ON THE PRICE OF THE HOMES IN CENTRAL PENNSYLVANIA 8 (Penn. State University Dep't of Architecture 2015), https://perma.cc/3AJ4-22CZ.
15   Pinecrest, FL, Code of Ordinances Ch. 30, Art., 5 Div. 5.27(a)(3)(a) (2018).
16   *Id.* Ch. 30, Art. 5, Div. 5.27(a)(3)(a)(1-2).
17   *Id.* Ch. 30, Art. 5, Div. 5.27(a)(3)(b)(1).
18   *Id.* Ch. 30, Art. 5, Div. 5.27(a)(3)(b)(1)(a).
19   *Id.* Ch. 30, Art. 5, Div. 5.27(a)(3)(c)(1)(c).
20   *Id.* Ch. 30, Art. 5, Div. 5.27(a)(3)(b)(1)(e).
21   City of El Paso, TX, Code of Ordinances §§ 18.10.366, 18.32.010 (2016).
22   INT'L CODE COUNCIL, 2015 INT'L RESIDENTIAL CODE, App. U § U103.1 (2016), https://perma.cc/QKD6-JH6K.
23   *Id.* § U103.4.
24   *Id.* § U103.6.
25   *Id.* § U103.7.
26   COUNCIL OF AMERICAN BUILDING OFFICIALS, RECOMMENDED REQUIREMENTS TO CODE OFFICIALS FOR SOLAR HEATING, COOLING AND HOT WATER SYSTEMS 19 (1980), https://perma.cc/B599-PRFG.
27   *Id.*

THIRD-PARTY CERTIFICATION REQUIREMENTS

3rd PARTY

REQUIRED

Kerrigan Owens (author)
Jonathan Rosenbloom & Christopher Duerksen (editors)

INTRODUCTION

There are several third-party certification programs for buildings. Each program measures different, but often overlapping, aspects relevant to building sustainably. Several of the programs include Leadership in Energy and Environmental Design (LEED), the Living Building Challenge, and Green Globes.[1] Each program establishes criteria that must be met before getting certified at several different levels.[2] LEED, for example, offers certification for buildings as Certified, Silver, Gold, and Platinum, and Living Building Challenge offers certification for buildings as Living Building Certification, Petal Certification, and Zero Energy Certification.[3] For each level of certification, the building must accumulate a certain number of points or meet other criteria to move to the next bracket.[4] Criteria may cover issues such as stormwater management, energy efficiency, or storage and collection of recyclables.[5]

In enacting these ordinances, local governments have a variety of options. They may make third-party certification a requirement or provide incentives to developers that achieve a certain level of certification. They may also adopt a carrot/stick approach by requiring a certain lower level of third-party certification, while also creating incentives to developers achieving a higher level. In addition, in drafting these ordinances local governments may select a variety of building sizes and types that are subject to the third-party certification requirements and when those buildings are subject to the requirements, such as upon sale, upon major renovation, or upon new construction.

EFFECTS

Buildings in the U.S. contributed to 13% of total water consumption and 38.9% of total carbon dioxide emissions. In addition, those buildings used 38.9% of total energy consumption.[6] Any ordinance incorporating third-party certification may provide a variety of economic, environmental, and social benefits.[7] For example, LEED buildings have been shown to decrease environmental impacts by integrating green design into the buildings and to improve the indoor environmental quality to increase productivity and comfort.[8] The minimum requirements for third-party certifications can also help save energy, water, and generate less waste.[9] Environmentally, depending on the certification being sought, the impacts can include improving air and water quality, conserving natural resources, and protecting biodiversity and ecosystems.[10]

By requiring new construction projects to comply with third-party certification minimums, communities may also experience economic benefits.[11] According to USGBC, LEED certified buildings cost less to operate, as they use more renewable energy sources, conserve electricity when necessary, and are longer-lasting due to the resilience of materials used for construction.[12] The World Green Building Trends 2016 Report found that on average, "Green Buildings" are 14% less costly to operate versus non-green buildings.[13] Further, green buildings are able to charge up to 30% higher rental rates.[14] A study by The World Building Council indicated that the higher the third-party certification, the higher the per square foot rental value of the building.[15] In addition, buildings that have third-party certifications for sustainability have a sale premium of 10-30 percent.[16] Further, the occupancy of third-party certified buildings is 23% higher than those without a green certification.[17]

EXAMPLES

San Anselmo, CA

To promote public health and reduce greenhouse gas emissions, the City of San Anselmo adopted an ordinance requiring LEED certification for certain new construction and renovations.[18] The ordinance requires that developers receive certain certification levels based upon the project's size.[19] New construction between 2,000 and 4,999 square feet must meet LEED certification.[20] Projects between 5,000 and 49,999 square feet must meet LEED Silver certification.[21] Finally, new projects over 50,000 square feet must meet LEED Gold certification.[22]

The ordinance requires renovations between 5,000 and 24,999 square feet to meet LEED Existing Building certifications and any renovations above 25,000 square feet to meet LEED Silver certification.[23] The ordinance includes additional requirements for sustainability, such as requiring photovoltaic pre-wiring and plumbing designs for future installation of solar water heaters.[24] The ordinance also allows for a "hardship or infeasibility" exemption in which applicants who believe there will be an infeasibility or hardship to comply with the certification levels can apply to be exempted.[25] The application must show that the project will continue to comply with the California Building Energy Efficiency Standards, and the building official will then determine a reasonable compliance for the project based upon the application.[26]

ADDITIONAL EXAMPLES

Boston, MA § 37 (2007) (stating that all proposed new construction subject to "Large Project Review" must meet LEED Certified).

Bonita Springs, FL § 10-126 (2012) (requiring all new residential permitted projects must meet the LEED Program for Homes standards, the current Green Home Designation, or the current Energy Start Certified New Home requirements).

Pittsburgh, PA § 909.01.M.3 (2004) (granting a higher density percentage of public projects if LEED certification levels are met. There is a 5% increase of density for Certified level, 10% for Silver level, 15% for Gold level, 20% for Platinum level).

To view the provision see San Anselmo, CA, Code of Ordinances § 9-19.010 (2010).

Miami Beach, FL

Miami Beach, Florida requires developers of new construction over 7,000 square feet to pay a fee of up to 5% of the construction cost.[27] The City states that the purpose for this program is to combat rising sea levels and mitigate climate change.[28] As a coastal city, Miami Beach views these goals as a priority for the City.[29] The City allows for a refund of the fee based on the level of sustainability certification received.[30] The City will refund 50% of the fee for projects that meet LEED Certified,[31] 66% of the fee for projects that meet LEED Silver , and 100% of the fee for projects that meet LEED Gold or Platinum.[32]

The fee goes to the City's sustainability fund which is used for environmental restoration projects, monitoring, green infrastructure, and stormwater quality improvement.[33] The sustainability fee program works as a carrot/stick approach to offer incentives for projects that meet higher levels of green building certifications while giving builders the option to not seek third-party certification and pay a fee.[34]

To view the provision see Miami Beach, FL, Code of Ordinances § 133-5 (2017).

ENDNOTES

1   *Leadership in Energy and Environmental Design (LEED),* U.S. Green Building Council, https://perma.cc/Y9BF-LRXD (last visited May 30, 2018); *Green Globes,* Green Building Initiative, https://perma.cc/RYD8-8LZB (last visited May 30, 2018); *Living Building Challenge,* International Living Future Institute, https://perma.cc/G6QG-TJ8S (last visited May 30, 2018).

2   *LEED, supra* note 1; *Green Globes, supra* note 1; *Living Building Challenge, supra* note 1.

3   *LEED, supra* note 1; Living Building Challenge, *Certification Pathways,* https://perma.cc/UT9H-A2XQ (last visited June 2, 2018).

4   *LEED, supra* note 1.

5   *LEED, supra* note 1.

6   *Buildings and their Impact on the Environment: A Statistical Summary,* Environmental Protection Agency (April 22, 2009), https://perma.cc/332S-BWEM.

7   *LEED, supra* note 1.

8   LEED, *supra* note 1.

9   *See LEED, supra* note 1; *see also Living Building Challenge, supra* note 1; *Green Globes, supra* note 1.

10 *Living Building Challenge, supra* note 1.

11 *See LEED, supra* note 1; *see also Living Building Challenge, supra* note 1.

12 *LEED, supra* note 1.

13 Anica Landreneau, *Green Buildings Don't Have to Cost More,* BUILDING DESIGN + CONSTRUCTION (May 02, 2017), https://perma.cc/2J6G-G6QV.

14 Jessica Dailey, *An Introduction to the Cost Benefits of Green Building,* CURBED UNIVERSITY (May 7, 2013), https://perma.cc/Y5V4-DCW6.

15 *Id.*

16 *Bricks, Mortar, and Carbon: How Sustainable Buildings Drive Real Estate Value,* Morgan Stanley Institute for Sustainable Investing (2016), https://perma.cc/EC6C-NSXL.

17 Daily, *supra* note 14.

18 San Anselmo, CA, Code of Ordinances § 9-19.010 (2010).

19 *Id.*

20 *Id* §9-19.040.

21 *Id.*

22 *Id.*

23 *Id.*

24 *Id.* §9-1.209, §9-1.210.

25 *Id.* §9-19.070(b).

26 *Id.*

27 Miami Beach, FL, Code of Ordinances § 133-6(a)(2016).

28 *Id.* § 133-2.

29 *Id.*

30 *Id.* § 133-6(a).

31 *Id.*

32 *Id.*

33 *Id.* §133-8(a), (c).

34 *See id.* § 133-6.

TREE CANOPY COVER

Alec LeSher *(author)*
Jonathan Rosenbloom & Christopher Duerksen *(editors)*

INTRODUCTION

A local government's tree canopy is the jurisdiction's area that is shaded by trees. Typically, as land is developed, the tree canopy is reduced because trees are removed to clear space for development. One study estimates that urban areas across the United States lost 36 million trees per year from 2009 to 2014.[1] Tree canopies provide numerous public and private benefits, including reduced air pollution, reduced heating and cooling demands, increased property values, improved physical and mental health, and reduced storm water runoff.[2]

This ordinance facilitates the growth of local tree canopy cover by requiring minimal tree canopy coverage per site or development, reforestation standards, and/or landscaping credits to developers that voluntarily plant more trees than required. Local governments have a variety of options when it comes to drafting these ordinances. They may set canopy minimums by percentage or area, may make the minimums applicable to residential, commercial, and/or industrial uses, and may set different minimums for different lot or development sizes. In addition, local governments may take a carrot and/or stick approach in which they require minimum standards and create incentives for those projects that exceed the minimums.

EFFECTS

Municipalities have much to gain by increasing the tree canopy. Carbon dioxide is one of the most prevalent and damaging greenhouse gasses (GHGs) that contributes to global warming.[3] Trees are a carbon dioxide "sink," meaning they naturally absorb carbon dioxide and release oxygen, intercepting that carbon dioxide before it enters the atmosphere.[4] Trees also provide shade that, when planted correctly, can help reduce the energy needed to cool a building during the summer and can help reduce the heat island effect.[5]

Likewise, trees can deflect harsh winds, saving energy on heating buildings in the winter.[6] By reducing energy needs, energy providers burn less fossil fuel and thus reduce GHG emissions.[7]

One study found that the presence of trees on a property increased the value by approximately seven thousand dollars.[8] This increased value is then used to increase local revenues through higher property tax assessments. Another study found that streets covered by shade from trees reduced the maintenance costs of the street by over 50% and increased street longevity by an estimated ten years.[9] Other studies have found that increased presence of trees benefits both mental and physical health.[10] For example, living near more trees reduces the risk of childhood asthma.[11] Finally, expanding tree canopy coverage provides more habitats for wildlife.[12] Thus, providing incentives to developers to bolster tree canopy is a valuable strategy for increasing all aspects of sustainability.

Developers also have much to gain by taking advantage of incentives for planting more trees. Some local governments grant increased density bonuses for developers that exceed the minimum required tree coverage.[13] This allows developers to create more profit producing units while simultaneously increasing the property's value.[14] Other municipalities decrease the minimum lot size requirement for developers that dedicate a tract exclusively to trees.[15] This allows developers to construct additional lots while providing greater tree canopy cover.

EXAMPLES

Charlotte, NC

To increase the citywide tree canopy cover to 50% by the year 2050, Charlotte requires a tree protection plan to accompany any application for grading, building, change of use, and zoning.[16] The plan must contain a root protection plan for any tree over two inches in diameter.[17] Charlotte also uses "tree save areas," which are areas in which an existing tree canopy exists that can be measured in square feet.[18] For residential developments, a minimum of 10% of the lot must be dedicated to a tree save area. For commercial developments, 15% of the lot must be dedicated to a tree save area. No building can be erected within ten feet of the edge of any tree save area. Developers are prevented from disturbing tree save areas unless the city grants a permit to do so.[19] Even if a permit is granted, the city may require the developer to "mitigate" the loss.[20]

If a developer fails to plant the required number of trees, a $50 fine per tree is assessed. Each day constitutes a new violation until the required planting occurs, up to a maximum fine of $1,000.[21] Further, if a developer causes damage that results in the total loss of a tree, s/he will be liable for the market value of that tree, up to a maximum of $20,000.

In a carrot and stick type approach, Charlotte also creates a number of incentives for developers who contribute more than they are required to the tree canopy. To encourage preservation of existing trees, a developer can seek an exemption from additional planting requirements if the developer preserves existing "heritage" trees. Heritage trees are trees listed on the *North Carolina Big Trees list*.[22] Residential properties with an area saving existing trees receive setback reductions.[23] Developers who dedicate a tree save area to a common open space can receive density bonuses if the titleholder covenants to maintain the area as a common, open space.[24] Additionally, some developments may qualify for lot width reductions if more than 25 percent of the lot consists of tree save areas.[25]

To view the provision see Charlotte, NC, Code of Ordinances §§ 21-91 to 21-126 (2010).

> **ADDITIONAL EXAMPLES**
>
> Ventura Cty., CA, Code of Ordinances § 8178-7.6.1 (2016) (requiring developers to plant 10 protected trees for each protected tree removed during development).
>
> Erie, CO, Unified Development Code §10.6.2 (C) (9) (2017) (granting developers of commercial and multifamily residential properties a reduction in required parking spaces for the preservation of trees beyond what is required by law).
>
> Lake Forest Park, WA, Municipal Code §§ 16.14.070 (2017) (establishing canopy coverage goals for different types of properties which are used in determining whether a tree removal permit will be granted).
>
> Fort Worth, TX., Code of Ordinances, App. A: Zoning Regulations § 6.302 (2009) (raising the city's canopy cover to 30% by requiring minimal levels of canopy cover on developments, but reducing the requirements if trees are planted elsewhere).

Baltimore, MD

After determining that its existing tree canopy provided only twenty seven percent coverage, Baltimore implemented a forest conservation plan aimed at expanding its tree canopy coverage to forty percent by 2030.[26] To that end, the city requires that every development over 40,000 square feet provide a forest preservation plan along with the associated building permit application.[27] The plan must contain a map of the land and identify the location and species of trees contained thereon, a three-year maintenance plan, an afforestation or reforestation plan, and other technical requirements.[28]

Baltimore also establishes an afforestation requirement, which requires developers to plant trees where there was previously no tree cover.[29] For lower density uses like agriculture and medium density residential developments, at least twenty percent of the land must be afforested.[30] For higher density developments like planned unit developments and high-density residential developments, only fifteen percent of the land must be afforested.[31] Developers are also required to retain existing trees that are associated with a historical site, or that are in a sensitive area such as a critical wildlife habitat or a wetland.[32]

To enforce these provisions, the city may assess a monetary penalty of $1.20 per square foot of area that is out of compliance.[33] That money is then put into the Baltimore County Forest Conservation Fund and may be spent on projects related to forest preservation.[34] Alternatively, the city may seek an injunction, preventing the developer from continuing construction until the violation ceases.[35]

To view the provision see Baltimore, MD, Code of Ordinances §§ 33-6-101 to 33-6-122 (2004).

ENDNOTES

1   David J. Nowak & Eric J. Greenfield, *Tree and Impervious Cover Change in U.S. Cities*, 32 URBAN FORESTRY & URBAN GREENING 32 (2018), https://perma.cc/3CX4-JVY2.

2   *Id.*

3   U.S. EPA, *Overview of Greenhouse Gasses*, https://perma.cc/7QKM-KKVE (last visited May 21, 2018).

4   Urban Forestry Network, *Trees Improve our Air Quality*, https://perma.cc/VV6E-KWLD (last visited May, 21 2018).

5   Hannah Safford et al., *Urban Forests and Climate Change*, United States Forest Service, https://perma.cc/QYA8-QUJM (last visited May 22, 2018).

6   *Id.*

7   Urban Forestry Network, *supra* note 4.

8   Geoffrey H. Donovan & David T. Butry, *Trees in the City: Valuing Street Trees in Portland, Oregon*, 94 LANDSCAPE & URBAN PLANNING 82 (2010), https://perma.cc/P9ZG-WTSX.

9   E. Gregory McPherson & Jules Munchnick, *Effects of Street Tree Shade on Asphalt Concrete Pavement Performance*, 31 J. ARBORICULTURE 303, 307 (2005), https://perma.cc/R3EZ-TD8B.

10   Omid Kardan et al., *Neighborhood Greenspace and Health in a Large Urban Center*, 5 SCI. REP. 11610 (2015), doi:10.1038/srep11610; Richard Ryan et al., *Vitalizing Effects of Being Outdoors and in Nature*, 30 J. ENVT'L PSYCH. 159, 167 (Nov. 2009), https://perma.cc/H5NE-7UJF.

11   G. S. Lovasi et al., *Children Living in Areas with More Street Trees Have Lower Prevalence of Asthma*, 62 J. OF EPIDEMIOLOGY & HEALTH 647 (May 2008).

12 *See e.g.*, Fred Sharpe, *The Biologically Significant Attributes of Forest Canopies to Small Birds*, 70 NORTH-WEST SCI. 86 (Jan. 1996), https://perma.cc/UQP3-MBMS.

13 *See e.g.*, Charlotte, NC, Code of Ordinances §21-95 (f) (2) (2010).

14 *See* Safford et al., *supra* note 5.

15 *See* Snohomish County, WA, County Code § 30.25.016 (h) (2010).

16 Charlotte, NC, Code of Ordinances § 21-92 (2010).

17 *Id.*

18 *Id.* § 21-2.

19 *Id.* § 21-93 (b).

20 *Id.*

21 *Id.* § 21-124 (b).

22 *Id.* § 21-95 (b); North Carolina Forest Service, *North Carolina Champion Big Tree Specimen Tree List*, https://perma.cc/9WLD-SJEN.

23 Charlotte, NC, Code of Ordinances § 21-95 (f) (1) (2010).

24 *Id.* § 21-95 (f) (2).

25 *Id.* § 21-95 (f) (3).

26 Scott Dance, *Baltimore's Tree Canopy is Growing, Slightly*, BALTIMORE SUN, Sept. 22, 2017, at https://perma.cc/A6PJ-ZHWR.

27 Baltimore, MD, Code of Ordinances § 33-6-103 (2004).

28 *Id.* § 33-6-108.

29 *Id.* § 33-6-111.

30 *Id.* § 33-6-111 (a) (1) (i).

31 *Id.* § 33-6-111 (a) (1) (ii).

32 *Id.* § 33-6-111 (b) (1); *Id.* at § 33-6-111 (b) (4).

33 *Id.* § 33-6-119.

34 *Id.*

35 *Id.* § 33-6-120.

URBAN GROWTH AREA

Alec LeSher *(author)*
Jonathan Rosenbloom & Christopher Duerksen (editors)

INTRODUCTION

Urban Growth Areas are designated areas in which certain development is permitted. Ordinances creating Urban Growth Areas often set legal boundaries separating urban, developable land from rural, conservation, preservation, non-developable land.[1] Any development outside of the Urban Growth Area is typically prohibited or greatly restricted. Urban Growth Area ordinances allow local governments to expand or contract growth areas upon certain findings, such as a finding of public welfare. Urban Growth Area ordinances may also contain provisions related to which entity has the power to modify the Urban Growth Area boundary, the procedure for reviewing the boundary, and how a developer may apply to build outside of the boundary. The local government may also set forth additional requirements for expanding the boundary based on the existing adjacent district. For instance, if a developer wants to build outside of the boundary and next to a district that is specifically designated for scenic views, the municipality may require a higher showing of necessity to develop than if the district was an agricultural zone. Finally, some ordinances have a sunset provision, by which the Urban Growth Area boundary will cease to exist unless the provisions are renewed.

EFFECTS

By creating an Urban Growth Area a local government is able to increase density and preserve natural resources, while not slowing growth. This prevents or limits urban sprawl and the issues associated with sprawl, including loss of support for public amenities, increased time spent travelling, degradation of water quality, and loss of habitat.[2]

Creating Urban Growth Areas helps lower public infrastructure costs by limiting expensive, underused infrastructure in low-density areas.[3] Providing public services such as water, sanitation, and transit services to larger geographic areas tends to be more expensive for the municipality.[4] Less dense areas also have higher public and individual transportation costs, as more fuel and roads (and associated maintenance) is required to move vehicles over the larger areas.[5] This has a direct economic cost and increases the amount of greenhouse gases produced by the cars and trucks dedicated to transportation.[6] Sprawl also contributes to increasingly expensive storm water management. As more land is developed and covered with impervious surfaces, the municipality must treat more polluted water runoff.[7]

Urban sprawl consumes vast areas of greenspace (land that is wholly or partially covered by vegetation) that were formally outside of the municipal area.[8] Greenspace is valuable for both recreational uses and ecological uses and purposes. Vegetation found in greenspaces consumes carbon dioxide thereby reducing the amount of carbon dioxide in the atmosphere that causes the greenhouse effect.[9] Removing greenspace reduces numerous ecosystem services that were being provided by the greenspace. Those services may include air and water purification, climate regulation, soil retention, habitat protection, and pollination.[10]

EXAMPLES

Pitkin County, CO

Pitkin County is home to the City of Aspen, and includes the rural, largely undeveloped areas surrounding Aspen. In response to neighboring rural counties being urbanized, the County promulgated comprehensive plans for each of the areas it serves and consolidated them into a single County comprehensive plan.[11] The common theme of the comprehensive plan is that it is intended to "sustain the existing rural character of Pitkin County."[12] The County achieved this goal by creating an Urban Growth Boundary, outside of which development is highly restricted, if not altogether prohibited.[13]

The County's Land Use Code codifies this requirement, providing that development will occur within the Urban Growth Boundary.[14] The Code does allow developers to propose building outside the boundary, but proposals are subject to a number of factors the Country must consider, most of which are averse to allowing such development.[15] Factors include requiring the new development to extend water and sewer utility lines and those lines will not be supported by the County; new development must show that its scale is such that development in the boundary is impractical; and new planned unit developments must minimize impacts to the environment.[16]

Developers must also contend with the County's policy of favoring preservation of agriculture.[17] Generally, the County discourages the fragmentation of agricultural land into smaller parcels, which is how most developers would acquire land beyond the boundary.[18] Further, the County generally does not expand municipal water and sewage systems outside of the Urban Growth Boundary, which places an additional burden of providing private systems on developers (see Establish Urban Service Area for more ordinances limiting utility services).[19]

To view the Comprehensive Plan see Pitkin Cty., CO, Pitkin County Comprehensive Plan (2003).

To view the Land Use Code provisions see Pitkin Cty., CO, Land Use Code tit. 8, ch. 1, § 1-50-30 (2016).

San Jose, CA

After experiencing rapid growth from 1950-1970, San Jose implemented an urban growth boundary (UGB) to contain urban sprawl.[20] The City had found that residential developments near the edge of city limits were not cost effective due to the substantial cost of expanding public services to new areas.[21] The City's general plan indicates that the City's goal is to use the UGB to minimize environmental impacts while insuring fiscal sustainability.[22]

The UGB is codified in the City's local planning ordinances. These ordinances generally require new development to occur within the existing UGB.[23] However, developers may build outside of the UGB if the City grants a modification to the boundary line location.[24] Modifications are classified as either "minor" or "significant." If a modification is minor, the application to change the UGB will be considered along with the annual review of the city's general plan.[25] Criteria for minor modifications include that the development is no larger than five acres (or twenty acres if the development creates a buffer that limits further development in the area), the development is immediately contiguous to existing developments in the UGB, and the development would provide an environmental benefit.[26] Notably, if a proposed development is contiguous to land that has already been granted a minor modification, then the proposal is automatically deemed significant.[27]

On the other hand, a significant modification is simply one that does not qualify as a minor modification.[28] Significant modifications are only considered when the city council finds that a review of the general plan is being scheduled for independent reasons, or that the applicant would be denied economically viable use of their land if the modification were not considered.[29] However, even if the city council makes one of these findings the modification would likely still be denied because the city code explicitly states that significant modifications are "strongly discouraged."[30]

To view the provisions see San Jose, CA, Code of Ordinances § 18.30 - 18.30.270 (current through 2018).

ENDNOTES

1   Myung-Jin Jun, *The Effects of Portland's Urban Growth Boundary on Urban Development Patterns and Commuting*, 2004 Urban Studies 1333, 1334; Portland Metro, OR, Ordinance 79-77.
2   James M. McElfish, Ten Things Wrong With Sprawl 1-5 (ELI 2007).
3   Daniel R. Mandelker, *Managing Space to Manage Growth*, 23 Wm & Mary Envtl. L. & Pol'y Rev. 801, 803.
4   *Id.* at 803.
5   *Id.* at 802.
6   *Id.* at 802.
7   Chan Yong Sun et al., *Impervious Surface Regulation and Urban Sprawl as its Unintended Consequence*, 32 Land Use Policy 317 (Dec. 2012), https://perma.cc/FYV4-JFMF.
8   *Id.*; Environmental Protection Agency, *What is Open Space/Green Space?*, https://perma.cc/L8FR-VVRS (last accessed May 18, 2018).
9   Byeongho Lee et al., *Carbon Dioxide Reduction Through Urban Green Space in the Case of Sejong City Master Plan*, in Proc. of the Int'l Conf. on Sustainable Bldg. Asia SB10 Seoul, at 538 (Feb. 2010), https://perma.cc/LE4C-DNT2 (last accessed May 18, 2018).
10   J.B. Ruhl, *The Twentieth Annual Lloyd K. Garrison Lecture: In Defense of Ecosystem Services*, 32 Pace Envtl. L. Rev. 306, 309 (2015).
11   Pitkin Cty., CO, Pitkin County Comprehensive Plan (2003).
12   *Id.* at 19
13   Pitkin Cty., CO, Land Use Code §§ 1-50-30, 1-60-60 (2016).
14   *Id.* §1-60-60.
15   *See id.* §§ 1-60-70, 1-60-80.
16   *See id.* § 1-60-70.
17   *See id.* § 1-60-80.
18   *Id.* § 1-60-80 (d).
19   *See id.* § 1-60-70 (d).
20   *Urban Growth Boundary*, City of San Jose, https://perma.cc/Y2BJ-Z774 (last accessed May 18, 2018).
21   *Id.*

22 *Envision San Jose 2040*, City of San Jose, https://perma.cc/AQ7R-KATC (Feb. 27, 2018).
23 San Jose, CA, Code of Ordinances § 18.30.100 (2018).
24 *Id.* § 18.30.100 (C).
25 *Id.* § 18.30.200 (B).
26 *Id.* § 18.30.300.
27 *Id.* § 18.30.300 (D) (2) (h).
28 *Id.* § 18.30.260.
29 *Id.* § 18.30.270 (B).
30 *Id.* § 18.30.270 (A).

URBAN SERVICE AREA

Alec LeSher *(author)*
Jonathan Rosenbloom & Christopher Duerksen (editors)

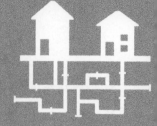

INTRODUCTION

An Urban Service Area (USA) is a defined area in which a municipality provides access to public services, such as water, sewer, and transit.[1] These areas are typically established in a local government's comprehensive plan and implemented in the zoning or other land use codes. Ordinances enforcing USA's effectively create a legal boundary, outside of which the local government is not obligated to provide public services and may refuse to do so. Developers are still permitted to construct beyond the boundary, but may be required to create their own connections to utilities and may also be refused the right to access those utilities. These ordinances should be structured in a way that allow the local government to expand or contract the USA to allow for more development or to restrain urban sprawl upon specified findings. In addition to establishing the boundary of a USA, these ordinances often set forth which services are restricted by the USA. Most commonly, this includes water, sewer, electrical, and transit services. Some municipalities also restrict the jurisdiction of the municipal police force to the boundary of the USA. These ordinances may set forth the entities responsible for reviewing the USA, procedures for modifying the USA, and sunset provisions on the applicability of the USA.

EFFECTS

An USA is a practical way of reducing urban sprawl and consequently increasing density by disincentivizing development outside of the USA. Urban sprawl is the process of a local government expanding into previously undeveloped land.[2] One of several issues associated with sprawl is the increased cost of public services in low-density areas.[3] Providing public services such as water, sanitation, and transit services to larger areas tends to be more expensive for the municipality and tends to pull resources from the urban core or previously developed areas.[4] Less dense areas also have higher transportation costs and a higher average of vehicle miles traveled.[5] This has a direct economic cost and increases the amount of greenhouse gases produced by the cars and trucks dedicated to transportation.[6]

Urban sprawl also consumes areas of greenspace (land that is wholly or partially covered by vegetation) that were formally outside of the municipal area.[7] Greenspace is valuable for both recreational uses and ecology. Greenspace acts as a carbon sink, trapping carbon dioxide before it enters the atmosphere, thereby reducing the greenhouse effect and associated climate changing impacts.[8] In addition, greenspace allows biodiversity to flourish and ecosystems to function more robustly. These ecosystems provide many critical services, such as the purification of air and water and soil retention.[9] USA's create an incentive for new developments to utilize unused area within the boundary, which cuts down on costs for the local government, and helps mitigate climate change.[10]

EXAMPLES

Baltimore County, MD

Baltimore County maintains an Urban Rural Demarcation Line (URDL).[11] This line marks the area in which public services are provided.[12] Although the ordinance does not prohibit construction, it limits public services and thus affects the practicality and expense of building beyond the URDL. Further, the zoning regulations restrict certain businesses and other uses to areas only within the URDL, limiting the type and level of activity in areas beyond the URDL.[13] The County further restricts growth within the URDL to specific growth areas to promote sustainable development.[14] For example, if a developer is intending to build a high-density apartment complex, they will be restricted from building outside the URDL, and even within the URDL the planning commission will likely encourage the developer to build in specific areas that have been specifically noted for additional growth.

To view the provision see Baltimore Cty., MD, Zoning Regulations §§ 101.1, 260.1, 430.3 (1979) (current through 2018).

To view the County Plan see Baltimore Cty., MD, Master Plan 2020, at 2 (2010).

Hillsborough County, FL

Hillsborough County adopted a USA in 1993 with the goal of 80% of growth occurring inside the USA boundary.[15] The current version can be found in the county's comprehensive plan.[16] As is typical of a USA, developers may still build outside of the boundary, although public services may not be readily available, if at all.[17] However, developments outside of the USA boundary can still apply for connection to public utilities.[18]

In the case of water treatment facilities, for example, the county uses a two-part test to determine if the county will create infrastructure to connect to new developments.[19] First,

the County Administrator must determine if the existing infrastructure has sufficient capacity to allow a new connection to the proposed development.[20] For this part, the County considers an engineering analysis of whether the existing public facilities can handle the extra burden created by the new development.[21] Second, the County Administrator must determine whether it would be feasible to connect public services to the new development.[22] This inquiry is largely a question of how far the development is from existing public utilities.[23] The farther away, the more costly such connection would be for the County, and thus the less likely that the connection would be deemed feasible. In this way, the two-part test deters developers from considering projects that are far from existing services and denser areas. Notably, developments within the USA boundary are deemed to satisfy this test without an inquiry into capacity or feasibility.[24] Thus, Hillsborough County has created some hurdles for developments outside the USA, while creating a strong incentive to develop inside the USA.

To view the provisions see Hillsborough Cty., FL, Code of Ordinances Part B § 102-67 to 102-68 (2000).

To view the county plan, see Hillsborough Cty., FL, Future of Hillsborough Comprehensive Plan for the Unincorporated Hillsborough County Florida – Future Land Use (2008).

ENDNOTES

1   New Hampshire Department of Environmental Services, *Urban Growth Boundary and Urban Service Districts*, https://perma.cc/GW34-SL8C (last visited May 24, 2018).
2   *Urban Sprawl*, Merriam-Webster, https://perma.cc/RQ52-DCTL (last visited May 23, 2018).
3   Daniel R. Mandelker, *Managing Space to Manage Growth*, 23 Wm & Mary Envtl. L. & Pol'y Rev. 801, 803.
4   Mandelker, *supra* note 3 at 803; James M. McElfish, Ten Things Wrong with Sprawl 2 (ELI 2007).
5   Mandelker, *supra* note 3 at 802.
6   Mandelker, *supra* note 4 at 802.
7   *Id.*; *What is Open Space/Green Space?*, U.S> EPA, https://perma.cc/ET63-53V6 (last visited May 18, 2018).
8   Byeongho Lee et al., *Carbon Dioxide Reduction Through Urban Green Space in the Case of Sejong City Master Plan*, at 538, https://perma.cc/3GY8-H6UX.
9   Keith Hirokawa, *Environmental Law from the Inside: Local Perspective, Local Potential*, 47 Envtl. L. Rep. 11048, 11058 (Dec. 2017); TEEB-Economics of Ecosystems & Biodiversity, TEEB Manual for Cities: Ecosystem Services in Urban Management 5 (2011), *available at* https://perma.cc/FJE4-H8X2; J.B. Ruhl, *The Twentieth Annual Lloyd K. Garrison Lecture: In Defense of Ecosystem Services*, 32 Pace Envtl. L. Rev. 306, 309 (2015).
10 Michael P. Johnson, *Environmental Impacts of Urban Sprawl: A Survey of the Literature and Proposed Research Agenda*, 33 Env't & Planning A 717, 721-22 (2001), https://perma.cc/L2UX-B3L3.
11 *Baltimore County's Land Management Areas*, Baltimore Co. Dep't of Planning, https://perma.cc/LCC2-8BZJ (last visited May 24, 2018); Baltimore Cty., MD, Zoning Regulations § 101.1 (1979) (current through 2018).
12 Baltimore Cty., MD, Zoning Regulations § 430.3 (1979) (current through 2018).
13 *Id.*
14 Baltimore Cty., MD, Master Plan 2020, 2 (2010).
15 *Urban Service Area - An Efficient Growth Management Tool*, Plan Hillsborough, https://perma.cc/D2PD-KWNJ (last visited May 24, 2018); Hillsborough Cty., FL, *Future of Hillsborough Comprehensive Plan for the Unincorporated Hillsborough County Florida - Future Land Use* § 1 (2008).
16 Hillsborough Cty., FL, *Future of Hillsborough Comprehensive Plan for the Unincorporated Hillsborough County Florida - Future Land Use* § 1.1 (2008).
17 *Id.* § 1.6.
18 Hillsborough Cty., FL, Code of Ordinances Part B § 102-67 (2000).
19 *Id.* § 102-68.
20 *Id.* § 102-68 (1).
21 *Id.*
22 *Id.* § 102-68 (2).
23 *Id.* § 102-68 (2) (a)-(c).
24 *Id.* § 102-68.

VEGETATION PROTECTION AREAS

Brandon Hanson (author)
Jonathan Rosenbloom & Christopher Duerksen (editors)

INTRODUCTION

Vegetation protection areas or zones designate sections of the land to be restored or kept as plant life. In vegetation protection areas local governments create zones that limit development and/or require the zone to only have plant life and, in some circumstances, native vegetation. Native plant life is beneficial not only for the environment but also for the community because native species require less water and maintenance, making the areas less expensive to maintain.[1] This ordinance should be drafted in a way that focuses on particular areas that have the most beneficial vegetation, which are often wildlife habitat areas.[2] A local government can create varying levels of protection in a vegetation protection area, such as primary and secondary areas. The primary vegetation area is fully protected while the secondary vegetation area establishes a variety of criteria to permit development.[3] Having varying levels of protection provides local governments with the flexibility to implement policies that are more tailored to their needs concerning native plants. Local governments can implement vegetation protection areas as part of a comprehensive green plan or as a standalone ordinance. Often a protection zone is located near wetlands or preserves to limit development in areas that are essential for wildlife habitats.

While this ordinance protects an area from being developed other ordinances help regulate based on species.[4] Protection areas are geared to protect critical species that are found in singular or multiple areas, while other ordinances in this Code are aimed at protecting existing vegetation, native species, and eradicating invasive species from the jurisdiction as a whole (See Require Native Trees and Removal of Invasive Trees; see also Require Use of Native Plants/Vegetation). In a vegetation protection zone the removal of vegetation for development purposes can be prohibited completely or limited. These zones are used to keep plant life in particular areas to ensure water filtration and necessary wildlife habitat.

EFFECTS

Overlay districts and zoning regulations concerning vegetation protection create ways to protect at risk areas from further harm. Protection areas allow wildlife to grow and move, ensuring greater biodiversity and helping property values.[5] Vegetation areas can increase land values and quality of life by safeguarding native species.[6] This ordinance can help preserve critical habitat for wildlife that other regulations are unable to fully protect.

Vegetation protection areas help reduce water consumption as native plants have adapted to the local environment, meaning they thrive on the precipitation common for the area and can handle other environmental stressors better than non-native plants (i.e. extreme temperatures, droughts, increased rainfall, etc.).[7] Further, traditional landscaping incorporating non-native species can hinder natural filtration processes. Studies show that infiltration rates are greater with native plant species than those without, and that areas with greater plant diversity have significantly higher rates of water interception than traditional turf landscaping.[8] Less water being used to sustain plant life helps keep water sheds healthy and has less of an impact on sewer and drainage systems. Along with a reduction on water consumption they promote soil health, provide critical habitat for local wildlife, and promote air quality.[9] The addition of vegetation protection areas can also help reduce greenhouse gas (GHG) emissions from lawn mowers and larger native species are more effective at absorbing GHGs and producing oxygen.[10]

EXAMPLES

Wayland, MI

Wayland created two overlay vegetation protection zones. The first zone, the natural vegetation zone, includes all land within 35 feet of the high-water mark of the river.[11] The natural vegetation zones are to have minimum development in the designated area.[12] The second zone is the transition zone, which extends for 15 feet beyond the natural vegetation zone.[13] The second zone is to provide distance between developed land and the natural vegetation zone. While the natural vegetation zone is intended to create a habitat for different species along the river, it is also used as a natural filtration system to reduce erosion and increase river bank health and maintain proper river temperature.[14] Site plans, land division plans, subdivision plans, condominium plans, unit development and all building permits must include a statement of compliance with the vegetation protection zones, except as permitted by the City code.[15]

To view the provision see Wayland, MI, Code of Ordinances § 20-520 (2006).

Kenai Peninsula Borough, AK

The Borough limits the amount of landscaping that can take place in designated habitat protection areas.[16] The habitat protection district created by the local government includes all land within 50 feet of the waters. The designated areas have restrictions on building different structures, to help preserve the area. The tract of land that is regulated is created to help protect the shore lines, preserve wildlife habitat, reduce pollution, and prevent damage to wetlands and other ecosystems.[17] Another unique protection the habitat areas provide is a tax credit. The credit is granted to any habitat protection or habitat restoration projects that are within 150 feet of any salmon stream or lake.[18] The tax credit allows, "For the first three consecutive tax years following completion of the project, the borough will provide a credit of up to 50% of the land tax assessment or the cost incurred in the project, whichever is less."[19] The also provides a tax exemption program for any increase in property value that is caused by the habitat protection and/or restoration.[20]

To view the provision, see Kenai Peninsula Borough, AK Code of Ordinances §§ 21.18.010 - 21.18.145 (2010).

ADDITIONAL EXAMPLES

Thurston County, WA, Code of Ordinances § 23.36.060 (2018) (requiring 60% of trees within a vegetation protection area to be evergreen, native trees).

Mandeville, LA, Code of Ordinances, App. A § 9.2.5.7 (2014) (prohibiting removal of oak trees within certain zones without the proper permit).

Tampa, FL, Code of Ordinances § 27-287 (current through 2018) (creating an overlay district that is designated as the Upland Habitat protection map).

Shasta Lake, CA, Code of Ordinances § 17.14.010 - 17.14.060 (current through 2018) (creating a habitat protection district designation that can be placed on areas of the City that have important wildlife habitat characteristics).

ENDNOTES

1   Audubon, *Why native Plants Matter: Restoring Native Plant Habitat is Vital to Preserving Biodiversity*, https://perma.cc/2XC9-W5LE (last visited July 3, 2018).
2   Logan Planning Scheme 2015 - Biodiversity Areas Overlay Code, https://perma.cc/RD6V-CWJS (last accessed 14 January 2018).
3   *Id.*
4   *See* Matthew King, *Exotic Removal Incentives Approved,* ENVTL. TIMES (2003); City of Palm Beach Prohibited Invasive Non-Native Vegetation Removal Ordinance, Art 14-Environmental Standards, Appendix 12 Incentive Programs (citizens of Palm Beach are only liable for up for $500 of the cost of invasive species removal).
5   *Id.*
6   Alaska.org, https://perma.cc/9BG2-M5Y4 (last visited Jan. 14, 2018).
7   *Landscaping with Native Plants: Quantification of the Benefits of Native Landscaping Current Knowledge,* EPA (December 6-7, 2004), https://perma.cc/4ZX8-9UQ6.
8   BRENDAN DOUGHERT & DAN SHAW, SUMMARY OF FUNCTIONAL BENEFITS OF NATIVE PLANTS IN DESIGNED AND NATURAL LANDSCAPES 3-4, https://perma.cc/FK54-B7L7 (last visited July 3, 2018).
9   *Id.*
10 Audubon, *supra* note 1.
11 Wayland, MI, Code of Ordinances § 20-520 (a)(1) (2006).
12 *Id.* § 20-520.1.
13 *Id.* at (a)(2).
14 *Id.* at (a)(1).
15 *Id.* at (b).
16 Kenai Peninsula Borough, AK, Code of Ordinances §§ 21.18.010 - 21.18.145 (2010).
17 *Id.* § 21.18.020 (a).
18 *Id.*
19 *Id.*
20 *Id.*

WATER EFFICIENT LANDSCAPING

Alec LeSher *(author)*
Jonathan Rosenbloom & Christopher Duerksen *(editors)*

INTRODUCTION

Efficient landscaping ordinances combine several techniques to reduce costs and time in the maintenance of landscapes.[1] These ordinances can be implemented in areas where there is too much or too little water.[2] They can be drafted to address flooding challenges or to supplement irrigation needs in particularly dry areas.[3] The ordinances include techniques such as the use of native and climate adapted plants to reduce the need for additional irrigation and maintenance.[4] Ordinances addressing water flow issues through landscaping may either create incentives to promote particular design and implementation by offering landscaping credits, or require certain landscaping during the development and approval process.[5]

EFFECTS

Water consumption is a primary concern associated with many types of landscaping.[6] Traditional lawns require large amounts of water to remain healthy and ascetically pleasing.[7] The use of this water places strain on the water supply, especially in drier climates, and is very costly to both the consumer and the local government.[8] Placing too much demand on the water supply can result in depletion of groundwater reserves and an increase in costs to maintain municipal water supplies.[9] Some landscapes also require chemical products to maintain healthy appearances, such as fertilizers, pesticides, and herbicides.[10] Many of these products can run-off their area of application and contaminate both ground and surface water, adding to costs and damaging ecosystems.[11]

Providing water to customers is also an energy intensive process.[12] Addressing stormwater run-off and providing potable water are some of local government's largest uses of energy.[13] Greenhouse gas (GHG) emissions needed to provide water should be considered in light of the utility of hydrating grass lawns.[14] In addition, lawn equipment, such as lawn mowers, may consume fossil fuels adding GHG emissions.[15] Switching to efficient landscaping can reduce and eliminate many of the costs associated with traditional landscaping, while reducing GHG emissions.[16] Selecting native and climate-adapted plants can reduce the need for irrigation as these plants are suited for area-specific precipitation patters.[17] Native plants are those that occur naturally in a region in which they evolved.[18] Once established, native landscapes outcompete weeds and eliminate the need for the harmful herbicides and pesticides required to maintain traditional turf grass, and greatly reduce the amount of GHG emissions from law care.[19] Laws exempting native species from being classified as weeds[20] can allow homeowners and municipalities to implement cost effective native alternatives to traditional turf grass, as native plant landscapes are largely self-sufficient.[21] Finally, thriving vegetation can help capture GHG emissions and lower energy costs involved with heating or cooling adjacent buildings.[22]

EXAMPLES

Fort Lauderdale, FL

Fort Lauderdale requires new developments to submit landscaping plans as part of the development permit process.[23] Developer landscaping plans are required to follow specific criteria in regards to the plant species used in the landscaping.[24] Fort Lauderdale organizes plant species into lists and sets a limit to the percentage that those species can be used.[25] For example, Norfolk Island Pine, Indian Rosewood, and Silk Oak are listed on the same list that limits them to no more than 10% of the trees present on a site.[26]

The ordinance also establishes criteria for the application of xeriscaping.[27] Xeriscaping is the inclusion of landscaping methods that promote water conservation.[28] Landscaping credits are provided for the preservation of native trees present on the site.[29] This continues to encourage developers to take advantage of present trees and native vegetation in the landscaping as opposed to replacing the plant life with non-native and potentially water-intensive species.[30]

To view the provision see Fort Lauderdale, FL, Unified Land Development Code § 47-39.A.13 (2009).

CA Model Ordinance

The State of California publishes and maintains a model water efficient landscape ordinance for use by local governments. Multiple municipalities in the state have adopted water efficient landscaping ordinances based on this model.[31] A key feature of this model ordinance is the requirement for a Landscape Design Plan.[32] The plan requires developers to make plant selection choice that are native to California and water efficient.[33] The plan also prevents the use of turf grass on steep slopes and requires special consideration for fire prevention in areas prone to wildfire.[34] Other parts of the model ordinance encourage the use of graywater (waste water from baths, washing machines, sinks etc.) and recycled water systems for irrigation.[35] The model ordinance also recommends the integration of stormwater management practices, such as rain gardens, infiltration beds, and constructed wetlands.[36]

ADDITIONAL EXAMPLES

Aurora, CO, Building and Zoning Code § 146-1437 (2004) (allowing a buffer reduction for landscapes that use xeriscaping or efficient landscaping practices).

Leavenworth, KS, Code of Ordinances, App. E § 6.06 (A) (2016) (providing bonus landscape credit for preserved native vegetation).

Sanibel, FL, Code of Ordinances § 126-675 (d) (2006) (requiring that 75 percent of vegetation be native when a parcel is developed or redeveloped in certain zones).

Scottsdale, AZ, Code of Ordinances § 46-106 (1989) (requiring that a person obtain a permit before removing any protected native plant species or face a fine of up to $2,500 or up to six months imprisonment).

Albuquerque, NM, Albuquerque Water Utility Authority Xeriscape Rebate (providing a rebate on water utility bill if the customer replaces traditional landscaping with low water use xeriscape).

Tuscon, AZ, Unified Code of Development § 7.6.4 (2015) (requiring use of drought resistant vegetation for new developments, with some exceptions, but allowing properties a small "oasis" zone where non-drought resistant vegetation may be planted).

The model ordinance includes additional sections that should be considered when drafting a local ordinance including topics on public education, irrigation design, and water waste prevention.

To view the model ordinance see CA Code Regs. tit. 23 §§ 490-495 (1992) (current through 2018).

ENDNOTES

1   Tobias Leanne et al., Retrofitting Building to be Green and Energy-efficient, 41 (2009); Sarah Schindler, *Banning Lawns*, 82 Geo. Wash. L. Rev. 394, 414 (2014).
2   *See, e.g.,* Aurora, CO, Building and Zoning Code § 146-1437 (2017); Fort Lauderdale, FL, Unified Development Code § 47-39.A.13 (2018).
3   Leanne et al., *supra* note 1, at 41; Aurora, CO, Building and Zoning Code § 146-1437 (2017).
4   *See* Leanne et al., *supra* note 1, at 41; Aurora, CO, Building and Zoning Code § 146-1437.
5   *See, e.g.,* Fort Lauderdale, FL, Unified Land Development Code § 47-39.A.13 (2018); Leavenworth, KS, Code of Ordinances § 6.06 (A) (2018).
6   Avi Friedman, Fundamentals of Sustainable Dwellings 199-200 (2012); Schindler, *supra* note 1 at 408-09.
7   Friedman, *supra* note 6 at 199-200; Schindler, *supra* note 1 at 407-08.
8   *See* Friedman, *supra* note 6, at 199-200; Schindler, *supra* note 1 at 410-11.
9   Friedman, *supra* note 6 at 169-70; Schindler, *supra* note 1 at 408.
10 Friedman, *supra* note 6 at 200.
11 Friedman, *supra* note 6 at 200; Schindler, *supra* note 1 at 410-11.
12 Schindler, *supra* note 1 at 409; Environmental Protection Agency, Energy Efficiency in Water and Wastewater Facilities 1 (2013)
13 Schindler, *supra* note 1 at 409; Environmental Protection Agency, *supra* note 12 at 1.
14 Schindler, *supra* note 1 at 409; Environmental Protection Agency, *supra* note 12 at 1.
15 Friedman, *supra* note 6 at 200-01.
16 Schindler, *supra* note 1 at 414.
17 Friedman, *supra* note 6 at 211.
18 Environmental Protection Agency, Landscaping with Native Plants: Factsheet (2012), https://perma.cc/F354-VEJ7.
19 John Ingram, When Cities Grow Wild: Natural Landscaping from an Urban Planning Perspective, Natural Landscaping: A New Landscape Ethic?, Wild Ones Natural Landscapers (1999).
20 *See e.g.,* Lee's Summit, MO Code of Ordinances § 30-36 (2018) (exempting native vegetation from classification as a weed, which would have required the owner of the property to destroy the vegetation).
21 Environmental Protection Agency, *supra* note 18.
22 Friedman, *supra* note 6 at 19-21; Department of Energy, *Landscaping for Energy-Efficient Homes,* Energy.gov (last visited Aug. 31 2017).
23 Fort Lauderdale, FL, Unified Land Development Code § 47-39.A.13 (C) (2009).
24 *Id.* § 47-39.A.13 (F).
25 *Id.* § 47-39.A.13 (F) (4) (e).
26 *Id.*
27 *Id.* § 47-39.A.13 (F) (9).
28 *Id.* § 47-39.A.13 (B) (cc).
29 *Id.* § 47-39.A.13 (F) (4) (g).
30 *Id.*
31 *See, e.g.,* Menlo Park, CA, Menlo Park Municipal Code §§ 12.44.010 - 12.44.220 (2017); Sonoma County, CA, Code of Ordinances §§ 7D3-1 to 7D3-9 (2018); Davis, CA, Davis Municipal Code §§ 40.42.010 - 40.42.220 (2018); Redding, CA, Code of Ordinances §§ 16.70.010 - 16.70.080 (2018).
32 Cal. Code Regs. tit. 23 § 492.6 (1992).
33 *Id.*
34 *Id.*
35 *Id.* § 492.15.
36 *Id.* § 492.16.

ZERO NET ENERGY BUILDINGS

Brandon Hanson (author),

Jonathan Rosenbloom & Christopher Duerksen (editors)

INTRODUCTION

The burning of fossil fuels is the primary source of greenhouse gas (GHG) emissions in the United States.[1] With the production of energy amounting to 28% of the U.S. total GHG emissions,[2] reducing the amount of energy produced through fossil fuels can have a large impact on the mitigation of GHGs.[3] This proposal seeks such reductions by establishing zero net energy requirements. Zero net energy buildings seek to produce as much energy as they use through renewable resources, typically based on annual energy use and production. Net zero energy buildings also promote more efficient energy consumption habits, as a reduction in consumption correlates directly with the production rate. Local government ordinances addressing net zero energy buildings provide for a variety of energy production types, including solar, wind, and geothermal.[4]

This proposal establishes zero net energy requirements for commercial and residential buildings. Alternative proposals can focus on creating incentives to build renewable energy sources (see Promote Renewable Energy with Incentives and Property Tax Exemptions for Renewable Energy Systems) and reduce energy demands by requiring some level of renewable energy production that is short of zero net or work towards net positive energy buildings, buildings that create more renewable energy than they use. In addition, the ordinances may be drafted to establish a set requirement for zero net energy or provide incentives to encourage zero net energy.[5]

Whether adopting zero net, net positive, or a lesser energy framework, these ordinances create incentives or establish firm requirements to build renewable energy systems and reduce energy demands. These ordinances set forth renewable energy standards for various building types, including new commercial or residential buildings or significant renovation to such buildings. The ordinances can be drafted to help local governments meet GHG reduction goals.[6] They may also work well with proposals to reduce energy demand through improved energy efficiency.[7] Solar energy is often the primary focus of these ordinances.[8]

EFFECTS

The most direct benefit stemming from zero net energy ordinances is a reduction in fossil fuels and associated GHGs that contribute to global warming.[9] Encouraging the installation of renewable energy can also have significant long-term economic benefits for those paying utility fees associated with energy.[10] By reducing reliance on utility company energy, through renewable energy sources, such as solar, net zero ordinances can decrease fees for both commercial enterprises and residential landowners.[11] The reduction of utility fees and other factors, lower operating costs, which lead lenders to be more active investors in these types of buildings.[12] The adoption of zero net energy requirements or lesser energy standards may lead to a net positive energy buildings and ordinances, creating a grid system that could one day be free from fossil fuels. Additional benefits stemming from the use of renewable energy include improved air quality, job growth, and energy security.[13]

Common thought is buildings that rely on renewable energy sources are expensive to build and burdensome, but developers in Utah have found only a 5 -10% increase in initial cost, some even claim no additional cost.[14] Even if the initial cost is more the operating costs are lower, as reliance on traditional energy sources are eliminated, and resale values are higher.[15] Making them less burdensome than buildings, without renewable energy systems.

EXAMPLES

Lancaster, CA

Lancaster, California has been a leader in requiring renewable energy.[16] After a long history of supporting solar development, Lancaster amended its building code to require that new buildings be outfitted with a solar energy system.[17] New single-family homes must have solar energy systems that can produce two watts of power for every square foot of the home.[18] This requirement can be modified if the builder provides documentation that a smaller system is able to meet the zero net energy requirements. If a developer cannot comply with the solar standards, they may be able to meet the requirements through other means.[19] Additionally, Lancaster adopted the California Energy Code, part of the California Code of Regulations, which requires most new residential buildings to have a solar ready area.[20] This area need not be outfitted with solar panels, but must have the capacity to install panels at a later date.[21] Multifamily residences must have a solar ready area of at least 15 percent of the total roof area.[22] Combined, these policies help Lancaster move toward zero net energy buildings. A process that the whole state of California is moving towards, with the plan from the California Energy Commission.[23]

To view the provision see Lancaster, CA, Energy Code § 15.28.020 (2017).

To view the provision see Cal. Code Regs. tit. 24 § 6-110.10 (b) (2016).

San Francisco, CA

San Francisco requires new residential buildings to be installed

with solar energy systems.[24] San Francisco also requires that the type of solar collectors installed meets a minimum requirement for per foot energy production.[25] For example, the code requires photovoltaic solar panels to produce at least 10 watts per square foot of solar panels.[26] Similar to Lancaster, the San Francisco code integrates a portion of the California Code of Regulations (CCR)[27] that requires new buildings to have 15% of total roof area available for solar panels.[28] Developers look to the CCR and identify the appropriate section based on building type, for example, multifamily apartment building. The CCR then establishes the minimum percentage of total roof area that must be ready for solar panels.[29] Then, according to the city solar requirements, the developer would install solar panels within the requisite roof area.

To view the provision see San Francisco, Cal., Green Building Code § 4.201.2 (2017).

To view the provision see Cal. Code Regs. Tit. 24 § 6-110.10 (b) (2016).

ENDNOTES

1   INTERGOVERNMENTAL PANEL ON CLIMATE CHANGE, CLIMATE CHANGE 2014: SYNTHESIS REPORT 48-49.

2   U.S. Environmental Protection Agency, *Sources of Greenhouse Gas Emissions*, epa.gov, https://perma.cc/UA9T-9VMH (last visited May 15, 2018).

3   *Id.*

4   U.S. Dep't of Energy, *A Common Definition for Zero Energy Buildings* (Sept. 2015) https://perma.cc/PFT2-K3R2.

5   ANDREW E. DESSLER, INTRODUCTION TO MODERN CLIMATE CHANGE 185-86 (Cambridge Univ. Press 2012); PERE MIR-ARTIGUES & PABLO DEL RÍO, THE ECONOMICS AND POLICY OF SOLAR PHOTOVOLTAIC GENERATION 172-75 (2016).

6   San Francisco Office of the Mayor, *Mayor Lee Announces Bold New Target of 50 Percent Renewable Energy by 2020*, sfmayor.org, https://perma.cc/XM6W-HFYT (last visited Sept. 18 2017).

7   Boston, MA, Municipal Code § 7-2.1; Minneapolis, MN, Code of Ordinances § 549.220 (12); Columbia, MO, Code of Ordinances §§ 27-161 - 27-169.

8   DESSLER, *supra* note 5, at 172. *See, e.g.*, Lancaster, CA, Energy Code §15.28.020 (c); San Francisco, CA, Environment Code §4.201.2; Sebastopol, CA, Sebastopol City Code § 15.72.

9   INTERGOVERNMENTAL PANEL ON CLIMATE CHANGE, *supra* note 1, at 48-49.

10   NATIONAL RENEWABLE ENERGY LABORATORY, DOLLAR FROM SENSE: THE ECONOMIC BENEFITS OF RENEWABLE ENERGY 2-3 (Sept. 1997), https://www.nrel.gov/docs/legosti/fy97/20505.pdf.

11   *Id.*

12   Mike Sheridan, *Net-Zero-Energy Construction Is Becoming More Cost-Effective*, URBANLAND (May 3, 2018), https://perma.cc/H56G-MKX2.

13   *Solar Energy Development on Federal Lands: The Road to Consensus: Before the Subcomm. on Energy and Mineral Resources of the H. Comm. on Natural Resources*, 111th Cong. 29 (2009); MIR-ARTIGUES & Río, *supra* note 5, at 172-173.

14   Erica Evans, *This Utah Group Just Proved that Net Zero Energy Buildings Don't Have to be More Expensive*, DESERT NEWS UTAH (June 13, 2018), https://perma.cc/27B8-665C.

15   Jessica Dailey, *An Introduction to the Cost Benefits of Green Building*, CURBED UNIVERSITY (May 7, 2013), https://perma.cc/79JH-CSN7.

16   Felicity Barringer, *With Help from Nature, A Town Aims to Be a Solar Capital*, N.Y. TIMES, Apr. 9, 2013, available at https://environmentcalifornia.org/media/cae/help-nature-town-aims-be-solar-capital.

17   Lancaster, CA, Energy Code § 15.28.020 (c) (2017).

18   *Id.* § 15.28.020 (c)(1).

19   *Id.* § 15.28.020 (d).

20   *Id.* § 15.28.010; Cal. Code Regs. tit. 24 § 6-110.10 (b) (2016).

21   Cal. Code Regs. tit. 24 § 6-110.10 (b)(1).

22   *Id.* § 6-110.10 (b)(1)(1).

23   *Commercial Zero Net Energy Action Plan*, California Public Utilities Commission (Dec. 2017), https://perma.cc/9GM4-QHX6.

24   San Francisco, CA, Environment Code § 4.201.2 (2017).

25   *Id.* § 4.201.2 (c) (1).

26   *Id.* § 4.201.2 (c) (1).

27   *Id.* § 4.201.2.

28   Cal. Code Regs. Tit. 24 § 6-110.10 (b).

29   *Id.* § 6-110.10 (b)(1)(B).

APPENDIX: ALTERNATE CATEGORIES FOR RECOMMENDED LOCAL LEGISLATION

To further assist local communities' use of this book, below please find all 43 recommendations contained in this book listed under one of four categories: Increase Renewable Energy; Increase Efficiency; Increase GHG Sinks; and Maintenance of Renewables. Some recommendations are listed under more than one category. Page numbers are parovided in parenthesis.

Increase Renewable Energy
District Heating and Cooling Zones (13)
Height & Setbacks to Encourage Renewables (19)
Renewable Energy for Historic Buildings (35)
Solar Energy Systems and Wind Turbines by-Right (40)
Density Bonus for Installation of Solar Energy Systems (52)
Limiting Off Property Shading of Solar Energy Systems (70)
Priority Parking for Hybrid & Electric Vehicles (80)
Property Assessed Clean Energy Program (84)
Property Tax Exemptions for Renewable Energy Systems (91)
Renewable Energy with Incentives (104)
Energy Benchmarking, Auditing, and Upgrading (126)
Limit Solar Restrictions in HOAs and/or CC&Rs (135)
Site & Solar Orientation (166)
Solar-Ready Construction (171)
Zero Net Energy Buildings (202)

Increase Efficiency
Accessory Dwelling Units (2)
Cluster/Conservation Subdivision in Rural/Urban Area (7)
District Heating and Cooling Zones (13)
Live-Work Units (23)
Local Recycling Centers (27)
Mixed-Use (31)
Tiny Homes and Compact Living Spaces (45)
Energy and Water Efficiency (56)
Green Roofing (60)
Infill Development (66)

Pervious Cover Minimums and Incentives (75)
Recycle, Salvage and Reuse Building Materials (96)
Recycling in Multi-family and Commercial Buildings (100)
Transit-Oriented Development (108)
Varying Unit Sizes within Multi-Family and Mixed-Use Buildings (112)
Alternative Pedestrian Routes to Parking Areas, Neighborhoods, and Businesses (118)
Energy Benchmarking, Auditing, and Upgrading (126)
Green Zones (131)
Maximum Size of Single-Family Residences (139)
Parking In-Lieu Fees (152)
Parking Maximums (157)
Safe Routes (161)
Third-Party Certification Requirements (176)
Urban Growth Area (185)
Urban Service Area (190)
Water Efficient Landscaping (198)
Zero Net Energy Buildings (202)

Increase GHG Sinks
Cluster/Conservation Subdivision in Rural/Urban Area (7)
Green Roofing (60)
Pervious Cover Minimums and Incentives (75)
Green Zones (131)
Native Trees and Removal of Invasive Trees (144)
Open Space Impact Fees (148)
Third-Party Certification Requirements (176)
Tree Canopy Cover (180)
Urban Growth Area (185)
Vegetation Protection Areas (194)
Water Efficient Landscaping (198)

Maintenance of Renewables
Limiting Off Property Shading of Solar Energy Systems (70)
Limit Solar Restrictions in HOAs and/or CC&Rs (135)

Index

accessory buildings, 3, 41, 43, 119, 150
accessory dwelling units (ADUs), 2–5, 46–50
affordable housing, 2–4, 46–47, 109, 112, 114–15, 131
agriculture, 7, 9–10, 183, 187
air quality, 82–83, 92, 97, 132, 145, 149, 183, 195
automobiles, 31, 67, 108, 113, 115, 120, 158–59
   charging stations, 80–82, 86
   electric vehicles (EVs), 80–83, 86
   hybrid electric vehicles (HEVs), 80–82
benchmarking, 126–28
biking, 32, 123–24, 157, 161–64
biodiversity, 8, 67, 149, 191, 195
building entrances, 122
building height requirements, 132
building materials, 43, 97, 140
building orientation, 166
building permits, 45, 150
commercial buildings, 23, 33, 100–102, 127–29, 171
covenants, 40, 135–36
district energy systems (DESs), 13–17
easements, 16, 70, 72–73, 122
ecosystem services, 8, 12, 186, 188, 193
electricity, 15–16, 41, 56–57, 70, 83, 85, 92, 101, 105, 140, 171
electric vehicles (EVs), see automobiles
energy
   audits, 126, 128–29
   benchmarking, 87, 126–27, 129
   costs, 14, 87, 126
   green, 52–53, 135
   sprawl, 53, 55
energy efficiency (EE), 13–14, 22, 38–39, 57–58, 61, 83–84, 89, 129, 176
Energy Star, 58–59, 127–29
fossil fuels, 13, 20, 32, 53, 57, 76, 85, 92–93, 104–5, 127, 181, 199, 202–3